Lead Chemistry

Edited by Pipat Chooto

Published in London, United Kingdom

IntechOpen

Supporting open minds since 2005

Lead Chemistry
http://dx.doi.org/10.5772/intechopen.87761
Edited by Pipat Chooto

Contributors
Amra Bratovcic, Pavel Arkhipov, Yury Zaykov, Osmel La Llave León, Jose M Salas Pacheco, Rajendra Sukhadeorao Dongre, Yi Xu, Tian Wang, Jie Zhang

First published in London, United Kingdom, 2020 by IntechOpen
IntechOpen is the global imprint of INTECHOPEN LIMITED, registered in England and Wales, registration number: 11086078, 5 Princes Gate Court, London, SW7 2QJ, United Kingdom
Printed in Croatia

British Library Cataloguing-in-Publication Data
A catalogue record for this book is available from the British Library

Additional hard and PDF copies can be obtained from orders@intechopen.com

Lead Chemistry
Edited by Pipat Chooto
p. cm.
Print ISBN 978-1-83962-567-1
Online ISBN 978-1-83962-568-8
eBook (PDF) ISBN 978-1-83962-569-5

For EU product safety concerns:
IN TECH d.o.o., Prolaz Marije Krucifikse Kozulić 3, 51000 Rijeka, Croatia, info@intechopen.com or visit our website at intechopen.com.

Meet the editor

After receiving his BSc. Degree from the Prince of Songkla University, which has been the main place for his work since 1983, Pipat Chooto received his PhD. Degree from Texas A & M University with preliminary expertise in the area of electrochemistry. Cooperating with his PhD. student, Dr. Chalermpol Innuphat, he got initiatives in lead analysis with a surprising number of readings and citations for the publication. Later when his two chapters for IntechOpen books were published, he worked as an editor of the present book, Lead Chemistry.

Contents

Preface

Despite the fact that some readers might think that the title of this book is too broad, this book should attract all of those interested in learning more about lead, especially in terms of recent interests and facts.

The most important thing is that the health concern for lead needs to be emphasized in addition to the fact that lead is still continuously in use all over the world and environmental aspects have not been well recognized. It might be hard to believe that the effects of lead penetrate deep down into the details of the human metabolic system and the shocking thing is that more research is needed because some of the aspects have no scientific data to support.

To describe the behavior of lead in its compounds and apply them in various areas, understanding the properties of lead in details is required. This would allow the readers to recognize that understanding this specific topic requires a combination of various disciplines, especially physics and spectroscopy.

Another fascinating aspect is the way that lead can form a variety of nanoparticles that can be very useful in a wide range of applications, especially engineering and analysis.

To make the book much more complete, one chapter has been dedicated to general concepts to inform newcomers as well as remind researchers who are familiar with lead as a general review on the role of lead in a large number of applications and disciplines.

In brief, it is hoped that our long sacrifice and contribution in the form of "Lead Chemistry" will reward the readers with knowledge as inspirations to further provide the world with fruitful, meaningful, interesting and creative concepts about lead, which should lead to greater and greater discoveries in the future.

Thanks to all the authors who worked patiently with us to the point of completeness. Our appreciation is extended further to all personnel involved in the process of publishing the book despite the fact that there were unexpected situations which have turned out to be a wonderful opportunity for success in publishing this book and, of course, laying a wonderful foundation for a lot more creative books in a few years to come.

Pipat Chooto
Prince of Songkla University,
Thailand

Section 1

Health Issues of Lead

Chapter 1

Effects of Lead on Reproductive Health

Osmel La Llave León and José M. Salas Pacheco

Abstract

It has been documented that lead can cause a wide range of adverse reproductive outcomes. In men, lead can reduce the libido and affect spermatogenesis reducing the quality of sperm. Other effects in exposed men include disturbance of prostatic function and damage in serum testosterone. In pregnant women, lead can cross the placenta and impair the development of the fetus. Therefore, exposed women are at risk of suffering spontaneous abortion, premature delivery, gestational diabetes mellitus, pregnancy hypertension, preeclampsia, premature rupture of membranes, intrauterine growth restriction, low weight birth, and other pregnancy complications. In both men and women, lead has been associated with infertility. Harmful effects of this heavy metal have been observed even at low levels of exposure. Thus, exposure to lead remains a public health problem, especially for reproductive health. Some strategies should be considered to prevent harmful effects of lead on both male and female reproductive systems.

Keywords: lead exposure, blood lead levels, reproductive outcomes, pregnant complications, sperm quality, infertility, reproductive health

1. Introduction

Lead is one of the most dangerous toxic metals. This metal has no known beneficial function in the human body. In contrast, lead can impair every system of the human body and specially the renal, hematopoietic, neurological, and reproductive systems. Exposure to lead has been associated with a broad range of physiological, biochemical, and behavioral and harmful effects. There is evidence of several reproductive damages in humans exposed to lead. In women, lead exposure has been associated with spontaneous abortion [1], low birth weight [2], preterm delivery [3], fetal growth restriction [4], premature rupture of membranes [5], pregnancy hypertension [6], preeclampsia [7], and gestational diabetes [8]. Maternal blood lead has also been associated with a decrease in length of gestation [9].

With respect to men, exposure to inorganic lead has been linked to a decrease in some parameters of semen quality. Lead exposure has been considered to adversely affect spermatogenesis [10] and reduced fertility [10]. High lead concentrations in seminal plasma can reduce the sperm count [10]. Sperm motility and sperm morphology also can be affected by lead [11].

The present chapter focuses on the harmful effects of lead on reproductive health of both men and women, due to the importance to established preventive measures to protect the health of parents and children exposed to this toxic metal.

2. Lead exposure and male reproductive health

Exposure to lead has been associated to several reproductive dysfunctions in men, such as decreased libido, impairment of spermatogenesis, and chromosomal damage, among others. However, studies about the relationship between lead exposure and male reproductive damage have shown inconsistent results. Most of the studies have analyzed the relationship between blood lead and semen quality due to the correlation observed between semen lead and blood lead [12]. Some studies have reported reduction in sperm count, morphology, and motility in men exposed to lead [13].

The effects of lead on sperm quality have been frequently studied in occupationally exposed individuals. National Institute for Occupational Safety and Health (OSHA) recommends that blood lead levels (BLL) above 40 μg/dL require health intervention. Nevertheless, studies in men without occupational exposure also showed evidence of the effects of lead on fertility. In a prospective, double-blind study carried out to evaluate the impact of seminal plasma lead levels on fertility, seminal plasma lead below this threshold value was associated with adverse effects on in vitro fertilization rates. In this survey, semen donors who participated in an artificial insemination program were included. Sperm lead concentrations were also negatively correlated with mannose receptors and mannose-induced acrosome reactions, the two biomarkers of sperm function [14]. These results show that increased lead concentrations in semen can harm male fertility.

Although most studies on the relationship between lead and infertility have been carried out in occupationally exposed workers, alterations in semen concentration of lead have been also observed in men without occupational exposure, probably due to other sources of exposure such as environment and foods. In a prospective and randomized clinical study carried out in men from infertile couples without occupational exposure to lead, a negative correlation between semen lead concentration and sperm count was observed in semen samples collected after 3–5 days of abstinence [10]. These results provide evidence that lead from environment and diet can also affect semen quality and, therefore, male fertility.

Several studies have evaluated the effect of lead exposure on the endocrine system. In lead smelting workers without clinical symptoms of lead poisoning, a decrease in serum testosterone (T) and an increase in steroid-binding globulin (SGG) levels were observed [15–18]. It is considered that lead impairs the majority of the endocrine glands. The analysis of the effect of long-term exposure to lead on thyroid function in exposed workers showed a negative association with T4 and FT4, and the depressed thyroid function was especially observed when the exposure was the most intensive [18]. In a group of workers occupationally exposed to lead from three battery factories, concentrations of FSH and LH were higher in comparison with a control group of non-exposed men, which constitutes an indicator that lead exposure alters testicular function [19]. From the biochemical point of view, it is considered that lead first causes testicular damage, and long-term exposure alters the hypothalamic-pituitary axis [17, 18, 20].

However, the results on this topic cannot be considered conclusive. In a study of the relationship between lead exposure and sex hormone levels in 133 men who had worked, at least for 6 months, in a battery manufacturing plant, BLL was measured, and endocrine system function was assessed by measuring testosterone, free testosterone, follicle stimulated hormone (FSH), and luteinizing hormone (LH). Workers were classified into two groups based on OSHA BLL standard: with BLL lower than 40 μg/dl and those with BLL equal or higher than 40 μg/dL. Statistical analysis showed no significant association between blood lead concentrations (BLC) and the sex hormone values. The authors concluded that lead exposure is not related to changes in male hormone levels [21]. In contrast, the evaluation of sperm count, sperm morphology, and hormonal levels (LH and FSH) of individuals attended in an infertility clinic in Iran showed negative significant correlations with BLL, while no correlation between BLC and sperm morphology was found [22].

Despite some contradictory results, there is a growing concern about the harmful effects of lead on male fertility, semen quality, and hormonal levels [15, 22]. Experiments in animal models have demonstrated that lead contributes to decreased male reproductive function [23]. In humans, lead exposure has been also associated with male endocrine dysfunction [24]. It is considered that oxidative stress plays an important role on male infertility. Lower total antioxidant capacity (TAC) and vitamin E concentrations were observed in seminal plasma of infertile men in comparison with fertile subjects [25]. In addition, there were significant differences between compared groups in accumulation of malondialdehyde. Moreover, concentration of malondialdehyde negatively correlated with sperm motility and morphology. On the basis of these results, it is suggested that seminal antioxidants and blood antioxidants can be used as biomarkers of sperm quality.

The effect of lead on reproductive health may vary due to the length of exposure. Taking into account the above-mentioned points, in a cross-sectional study of male workers, the effects of current and long-term occupational lead exposures on several biomarkers of male reproductive health were evaluated [11]. Semen and blood samples from male employees of a lead smelter were obtained, and concentrations of testosterone, follicle stimulated hormone, luteinizing hormone, and blood lead were determined. A decreasing trend in total sperm count was observed in relation to the increase in BLL. In addition, total motile sperm count, sperm concentration, and total sperm count showed an inverse relationship with long-term lead exposure. Nevertheless, lead exposure was not associated to sperm motility, sperm morphology, or serum concentrations of reproductive hormones.

The effects of lead exposure on male reproductive function have also been studied in animals. Experiments in mouse have shown that lead can interfere with the binding of androgens [26], suppress follicle stimulating hormone production [27], affect the function of Sertoli cell, and increase the lactate production, which constitute an essential substrate for spermatogenesis [28]. Lead exposure has been also associated with decreases in the activity of testicular oxidizing enzymes [29] and in the synthesis of testicular RNA in rats [28]. A study conducted in rats showed a positive correlation between blood lead and levels of lead in epididymal sperm and demonstrated that lead can cause generation of reactive oxygen species in sperm, which led to oxidative stress and, therefore, impairment of sperm function [30].

Epidemiological data indicate that exposure to lead can cause prostate diseases in adult males. In a study, blood lead in patients suffering from prostate cancer (PCA), patients with benign prostate hyperplasia (BPH), and a control group of men living in similar socioeconomic conditions was examined [31]. Results indicated significant higher concentrations of lead in blood in PCA and BPH males in comparison with controls. In addition, patients with PCA and BPH had significantly lower blood levels of zinc and copper than the comparative group. It is well known that Zn has an essential role in the regulation of prostate epithelium homeostasis and in ejaculation [32]. Zinc is a cofactor for many enzymes and an essential metal for the integrity of cellular membrane [33]. Lead can displace zinc ions at the proteins, provoking the inhibition of the enzymes. The displacement of zinc by lead in seminal fluid could determine the effects of prostate function, leading to decreased fertility [32].

Some authors consider that the main effect of lead on the male reproductive system is the alteration of the reproductive hormonal axis and the hormonal control of spermatogenesis, instead of the direct effect on the seminiferous tubules of the testes [23, 34]. Moreover, there is evidence that the blood-testis barrier acts as a protection for the testis cells against the harmful effects of lead [35, 36]. On the other

Effects of lead on male reproductive system
➢ Decrease in sperm count
➢ Decrease in sperm motility
➢ Abnormal sperm morphology
➢ Decreased volume of ejaculation
➢ Decreased serum testosterone
➢ Increase in serum follicular staining hormone (FSH)
➢ Increase in serum luteinizing hormone (LH)
➢ Impairment of sperm DNA

Figure 1.
Some effects of lead on male reproductive system.

hand, some researchers pay more attention to the impairment of sperm parameters, such as volume of ejaculation, sperm density, abnormal morphology, sperm count, and motility, by the toxic effect of lead [10, 14, 22, 37].

Although the mechanisms by which lead affects male reproductive health are still unclear, there is no doubt that this toxic metal can jeopardize fertility in men due to alterations in semen quality, in the function of reproductive hormones, or both (see **Figure 1**). Despite conflicting reported results, there is growing evidence that lead exposure, even at low levels, can impair male reproductive health. Future research should deepen the analysis concerning these issues.

3. Lead exposure and female reproductive health

It is well known that lead has harmful effects on female reproductive system. Women at reproductive age are at risk of suffering some health disorders due to the toxic effects of this metal. Occupational exposure to lead is more frequent in men compared to women. However, there are some reports on the harmful effects of lead suffered by women who work in places where lead or some lead compounds are used. In a study conducted to determine the effects of occupational exposure on bone and lead blood levels, women who were former workers at a smelter were compared with a cohort of women with no-known occupational exposure. Higher levels of lead in blood and tibia were found in the exposed group. In addition, the difference in bone lead levels between compared groups was significantly higher than the difference in BLCs [38]. In accord with these findings, a study carried out in Mexico showed that women who work with lead have greater probability to have BLCs above the CDC recommended value of 5 μg/dL compared to non-exposed women [39].

It is necessary to consider that women can be exposed to lead not only at work but also through the clothes, shoes, and work instruments that are taken home by the cohabitants who work in places where lead is used. Higher BLCs in pregnant women who live with someone who is exposed to lead at work in comparison with those who live in houses where nobody works in places that lead is used have been observed [40]. In addition, lead exposure may occur when women use some cosmetics, such as surma or kolh, and other beauty products [41–43].

Women can also be exposed to lead by pica habit, an eating disorder that consists of the consumption of non-food items without nutritional value. Among the most harmful types of pica is the consumption of soil, paint chips, and pottery. Pregnant women consuming these items put both themselves and the fetus at risk of lead poisoning [44, 45]. In Mexican, women who were recognized that they used to eat soil had significantly higher BLL compared to those who did not have this habit [40]. In one study in New York, pica behavior among lead-poisoned pregnant women (BLL $\geq$ 20 μg/dL) was 9%. The most common practice among them was eating soil (64.6%). The probability of having BLLs ≥40 μg/dL among women who reported pica was three times higher in comparison with those women who did not report this habit. In addition, pica-reporting women had a mean peak of BLL during pregnancy significantly higher compared to those who did not report pica (29.5 μg/dL vs. 23.8 μg/dL) [45].

In addition to the effects of lead on women's fertility, a wide range of published reports refers to the damage caused by this heavy metal during pregnancy (see **Figure 2**). Prenatal exposure to lead can cause several obstetric complications and adverse pregnancy outcomes [46]. Lead absorbed into the body, mainly by ingestion or inhalation, enters the bloodstream and accumulates in soft organs (mostly in brain, liver, and kidney) and bones [47, 48]. It is considered that lead in bone represents approximately 95% of the total body burden in adults [47]. During pregnancy, the demand of calcium rises, and lead stored in bone can replace the calcium and recirculate in the bloodstream, becoming an endogenous source of exposure [16, 48–50]. Lead from the blood can cross the placenta and impair the development of the fetus [51, 52]. Therefore, lead-exposed women are at risk of suffering various pregnancy complications, such as spontaneous abortion [1, 53], preterm delivery [54, 55], GDM [8, 56], pregnancy hypertension [57–59], preeclampsia [60–64], premature rupture of membranes [65, 66], intrauterine growth restriction [67], and low weight birth [68, 69], among others.

Although some researchers have failed to demonstrate the relationship between lead and abortion [70, 71], a study conducted in Mexico showed evidence that, even low-to-moderate lead exposure, below 30 μg/dL of blood lead can increase the risk of spontaneous abortion [1]. In this case, the range of BLLs in pregnant women was 1.4–29 μg/dL. Those lead concentrations can be considered common in general population in many countries, and lower to those observed in occupationally exposed women. It is considered that the mechanism by which lead induces abortion is related to the direct transmission of the metal to the developing fetus due to the demineralization of bones during pregnancy [72, 73].

Several studies have confirmed that pregnant women exposed to lead have more probability of having a preterm delivery compared with non-exposed women. Nevertheless, results are still inconsistent. In a prospective cohort study carried out in China, maternal urinary lead was measured and adjusted by creatinine, and newborns were classified as preterm birth and early term birth. The mean urinary lead levels were significantly higher in preterm births. In addition, among all newborns, an increase in maternal urinary lead was associated with a decrease in gestational age [3].

Effects of lead on female reproductive system
➢ Impairing menstruations
➢ Reduction of libido
➢ Premature rupture of membranes
➢ Intrauterine grow restriction
➢ Abortion
➢ Premature delivery
➢ Stillbirths
➢ Pregnancy hypertension
➢ Preeclampsia
➢ Neonatal death
➢ Gestational diabetes
➢ Low birth weight

Figure 2.
Harmful effects of lead on female reproductive health.

Lead can displace calcium because they both have similar chemical characteristics and follow analogous metabolic pathways [74]. It has been recognized that when lead crosses from the bloodstream to the placenta, the growth of the fetus can be impaired due to the interference of lead with calcium metabolism [68, 69]. The evaluation of prenatal exposure to lead has shown inverse association between maternal urine lead levels and preterm low birth weight [68]. Other studies analyzed the relationship between the levels of lead in tibia and patella and birth weight, considering that bone lead is a better biomarker to estimate the effect of lead on the fetus compared to blood lead [75, 76].

In a study conducted to evaluate the relationship between lead exposure and birth weight in Mexican women, lead levels were measured in maternal venous blood, umbilical cord, and tibia and patella. The weight of newborns was determined within the first 12 hours of delivery. Although all biomarkers of lead exposure were negatively associated with a decreased size of newborns, this association resulted statistically significant only for tibia lead levels [75]. Similar results were observed in the analysis of the relationship between maternal lead burden and early postnatal growth in a cohort of breastfed newborns [75]. In this study, maternal BLL measured at 1 month postpartum and maternal bone lead levels were significantly associated with infant BLL. Moreover, infant BLL and maternal patella lead level were inversely associated with weight gain. The weight gain from birth to the first month of life was 142 g lower in infants with BLL ≥ 10 compared to those with lower BLCs.

There is growing evidence that lead is a risk factor for gestational diabetes mellitus. Experiments with rats have demonstrated that lead exposure can induce glucose intolerance and hyperglycemia [8]. But epidemiological studies showed contradictory results. In women at 22–28 weeks of gestation, slightly mean BLCs were observed compared to those without GDM, but this difference was not statistically significant. The geometric mean BLCs were 6.13 ng/g in women with GDM and 6.05 ng/g in women without GDM. Based on this result, authors suggested that lead at these low levels of exposure is not associated with the risk of suffering GDM [77]. In contrast, in a French mother-child prospective cohort study, blood lead was associated with IGT, supporting the evidence that maternal exposure to lead is a risk factor for GDM [56]. Further studies have to be performed to confirm the deleterious effect of lead on metabolic processes and, particularly, on the development of GDM.

A large number of investigations provide evidence that exposure to lead is associated with hypertension in adults [78–81]. For this reason, the question of whether lead is associated with gestational hypertension (GH) and preeclampsia (PE) has gained a great importance in recent years.

In a cohort of pregnant women in Los Angeles, California, blood and bone lead were assessed in the 3rd trimester and post-delivery, and the prevalence of hypertension was measured [82]. The relationship of both biomarkers with GH was analyzed. After adjusting by covariables, no significant association between BLLs in 3rd trimester and hypertension was observed. Nevertheless, calcaneus bone lead was significantly associated with the risk of hypertension.

In a cross-sectional study with Maltese Caucasian women at third trimester of gestation, significantly higher BLCs in hypertensive women compared to normotensive women were observed [83]. Moreover, BLL showed a positive relationship with systolic and diastolic blood pressure.

The relationship between BLL at mid-pregnancy and blood pressure was assessed in a study carried out in pregnant women of two French municipalities [83]. In this study, hypertensive women had significantly higher BLL than normotensive women. Additionally, lowest frequency of hypertension was observed among women in the lowest quartile of BLL. These findings are in accord with those observed in Nigeria, in which the impact of lead on pregnancy outcomes was investigated [84]. Significantly higher frequency of hypertension was observed in women with BLL ≥ 10 μg/dL compared to those who had lower concentrations of lead in blood.

The findings on the association of lead exposure with GH led to investigate if this toxic metal could be considered a risk factor for preeclampsia, a pregnancy disorder characterized by high blood pressure and proteinuria detected after 20 weeks of gestation [85]. In a cross-sectional study that included women between 29 and 43 weeks of gestation, significantly higher concentrations of lead in red blood cells of pregnant women diagnosed with preeclampsia were found compared to those without hypertension. Furthermore, women with severe preeclampsia had also higher blood cell lead concentrations than mild preeclamptic women [61]. In contrast, in a case-control study conducted in women without occupational exposure, BLCs measured within 24 hours of delivery did not differ between women with preeclampsia and normotensive group, but a significant difference between the groups was observed with respect to umbilical cord lead (UCB) concentration [64]. In addition, the ratio of umbilical cord lead to whole blood lead was significantly associated with preeclampsia.

Despite the contradictory results of some studies, the majority of those supported the hypothesis that lead can cause preeclampsia. Some possible mechanisms have been suggested to explain the roll of lead in the development of this pregnancy disorder. It is considered that lead increases the circulating levels of endothelin, a vasoactive substance that causes constriction of the blood vessels, leading to the increase of blood pressure [63]. Lead also interferes in the increase of reactive oxygen species reducing the serum levels of nitric oxide (NO) and other vasodilator substances [86–89]. From the molecular point of view, lead causes inhibition of membrane adenosine triphosphatases (ATPases), which produces vasoconstriction due to the increase of intracellular calcium ions [63, 90].

4. Effects of gender on lead toxicity

The influence of sex in the effect of lead on health is a controversial subject. Although sex differences regarding exposure, absorption, and metabolism of lead have been reported by certain researchers [91, 92], the results are not conclusive. In a prospective cohort, the effects of gender differences in the relationship between lead exposure and neurodevelopmental toxicity were analyzed [91]. Lead levels were determined in maternal blood in early and late pregnancy, in cord blood at birth, and in children's blood at 2, 3, and 5 years old. As a result, significant association between lead concentrations at late pregnancy and the risk of behavioral problems was observed in males, while blood lead measured in 2- and 5-year-old children was associated with an increased risk of behavioral problems in females. According to previous data, early in life, the susceptibility to neurotoxic effect of lead is higher in boys than in girls. On the other hand, experimental data suggest that susceptibility to immunotoxic effects of lead is higher in females [92]. More research is needed to elucidate these inconsistencies.

The biological effects of lead exposure in human also appear to be different according to the gender. In a study carried out in Japan, the authors aimed to determine the effects of gender on porphyrin metabolic disorders induced by lead exposure [93]. Blood lead, plasma delta-aminolevulinic acid (ALA), urinary ALA, and urinary coproporphyrin (CP) were determined in exposed workers. Although no significant differences in blood lead concentrations between male and female workers were observed, women had higher plasma ALA concentrations, as well as higher excretion of urine ALA and CP in comparison with men. The mechanism that could explain this difference is still unclear.

With respect to the reproductive system, health damages in female have been observed even at very low levels of exposure. In a study carried out in Taiwan, the relationship between low-level lead exposure and risk of infertility was evaluated [94]. The average lead concentration in infertile women (3.5 μg/dL) was significantly higher than in a control group (2.78 μg/dL). Furthermore, women with BLL >2.5 μg/dL had a threefold higher risk of infertility than those with BLL $\leq$ 2.5 μg/dL. In contrast, the harmful effects of lead in male reproductive system have been detected at higher levels of exposure than those in female. The importance of finding explanation of gender effects for lead and other environmental toxic substances was discussed by the Society for Women's Health Research in a roundtable at the National Institute of Environmental Health Sciences in October 2002 [95].

Despite the above-mentioned results, gender differences in susceptibility to lead poisoning have been considered in few investigations. Some studies have included gender as a confounding factor in the relationship between lead exposure and health impairment [96]. However, in other investigations, differences between male and female in regard to the harmful effects of lead have not been found.

According to the results of the investigations, the following differences between men and women regarding lead exposure can be highlighted:

- Generally, in non-exposed individuals, blood lead levels are higher in males than in females.
- Damages to female reproductive health can occur at lower levels of exposure than in men.
- The risk of suffering behavioral problems in relation to prenatal lead exposure at early childhood is higher in females.
- The susceptibility to neurotoxic effects of lead appears to be higher in boys than in girls.
- The susceptibility to immunotoxic effects of lead is higher in females.
- From the biological point of view, porphyrin metabolic disorders induced by lead exposure affect females more than males.
- The impairment in the synthesis and function of hormones has been observed in both genders. But, little is known about the differences between male and female regarding the mechanisms by which lead affects the reproductive system.

In summary, gender difference should be considered an important factor for a better evaluation of the harmful effects of lead on health. Further research is needed to better understand the role of sex as a modifier of the effects of lead exposure.

5. Lead exposure and children's reproductive system

Lead is considered to be able to affect the development of children's reproductive system. There is evidence that both paternal and maternal lead exposure can cause a detrimental impact on the structure and function of gametes, which might cause adverse effects on newborn's health [97]. Embryos and fetus are extremely sensitive to environmental toxicants. Exposure to lead during pregnancy is known to be able to impair fetal development, since lead can cross the placental barrier and reach the fetus [51, 52].

Lead is known to affect testosterone levels in adults, leading to reproductive dysfunction. Low levels of testosterone can reduce semen quality in men and increase genital malformations [98]. In contrast, high levels of testosterone in women are associated with higher frequency of polycystic ovary syndrome (POS) [99] and puberty disorders [100]. In spite of this, there are few studies that have focused on the relationship between lead exposure and androgen hormone levels in children. One of the few longitudinal studies on this issue was carried out in Russia [101]. This study evaluated the impact of organochlorine chemicals and lead in growth and pubertal timing in 516 boys. Children were enrolled in the study at the age of 5–7 years and were followed up until the age of 18–19 years. Lead exposure was negatively associated with growth during puberty. In addition, it was suggested that lead may delay the timing of male puberty.

In a study carried out to evaluate the relationship between blood metal concentrations and testosterone levels in the USA, children and adolescents' concentrations of lead, cadmium, mercury, and selenium in blood, as well as serum testosterone levels, were determined [102]. Although no significant association between blood lead and total testosterone (TT) was observed, the concentrations of TT were significantly higher for girls in the fourth quartile compared to those in the first quartile. On the other hand, in a prospective study conducted in Mexico City, maternal patella lead and early childhood blood lead were inversely associated with breast growth in girls [103]. Furthermore, an increase in girl's maternal patella lead was associated with later age of menarche. In addition, blood lead during childhood negatively associated with pubic hair growth in girls. No associations were observed in boys.

6. Preventive strategies to avoid harmful effects of lead on reproductive health

The fact that lead exposure is related to a wide range of adverse effects on reproductive health is accepted by most researchers nowadays. The main source of exposure remains occupational. But there is no doubt that, in recent years, the environmental exposure to lead has decreased, especially in developed countries like the United State, Canada, and others [104]. A main role in this reduction is attributed to the elimination of leaded gasoline [105]. Nevertheless, the risk of lead poisoning still remains, mainly in developing countries, due to some sources of exposure, such as lead paint, cosmetics, traditional medicines, electronic waste, and glazed ceramic vessels, among others [106].

It is very difficult to dispense with lead due to its uses in a wide range of industrial lines, such as smelting, manufacturing and recycling of car batteries, and lead crystal glassware. However, in recent years, there has been an increase in the diffusion of the damage that lead can cause and the measures that must be taken to protect people's health. The identification of risk factors for having high BLLs has contributed to reduce the prevalence and severity of lead poisoning. In a way, the results of research have helped people to become aware of the toxicity of this metal and the danger it poses to their health.

Health interventions in last decades have led to a decrease in lead exposure. In spite of this, it is necessary to increase protection measures, especially for women and children. To date, there is no exposure lead level that can be considered safe. Although the CDC established 5 μg/dL as the reference value for BLLs in children, epidemiological research has demonstrated that even at lower lead concentrations adverse health effects can occur. With respect to females, adverse reproductive outcomes have been observed also at BLLs below 5 μg/dL, decrease in delta-aminolevulinic acid dehydratase (ALAD) activity has been detected in pregnant women at mean blood lead ≥2.2 μg/dL [107], and damages in female reproductive system have been reported at BLLs above 2.5 μg/dL [95]. Some prevention strategies should be considered for protection of the toxic effects of lead. Preventive measures should include at least the following:

- Developing public awareness campaigns to identify sources of health exposure
- Evaluation of risk factors for all pregnant women in their prenatal care
- Education of childbearing age women to avoid sources of lead exposure
- Screening of lead exposure for all pregnant women by means of diagnosis tests, such as blood lead, ALAD activity, and urine ALA
- Keeping children and pregnant women with BLL ≥ 5 μg/dL out of exposure sources
- Requiring employers to take measures to reduce lead levels in workplaces
- Requiring exposed workers to use protective means
- To carry out a review of regulations to ensure greater protection for the exposed population
- Increasing environmental monitoring (lead in air, soil, water, etc.) to detect any deviation from established standards
- Searching for new biomarkers, with high sensitivity and specificity, to assess the exposure and effects of lead in the body

7. Conclusions

There is enough evidence that lead exposure can harm reproductive health of both men and women. The harmful effects of lead have been mostly observed in occupationally exposed people. Nevertheless, in recent decades, research has demonstrated that these damages can occur at levels of lead formerly considered harmless. Most observed effects on male reproductive system are related to the direct impact of lead on semen quality, such as volume of ejaculation, sperm density, abnormal morphology, sperm count, and motility. In addition, lead can alter the concentrations of some male reproductive hormones, such as follicle stimulating hormones, testosterone, and luteinizing hormone.

In women, prenatal exposure to lead, even at very low levels of exposure, has shown to be harmful for both the mother and the fetus. Thus, any level of lead exposure could be associated with adverse reproductive outcomes. Lead has been associated with a wide range of adverse outcomes, including spontaneous abortion, intrauterine growth restriction, premature delivery, stillbirths, pregnancy hypertension, preeclampsia, and low birth weight, among others.

Several recent studies have suggested hypothesis related to the mechanisms by which lead affects male and female reproductive health. However, more research is needed to clarify these mechanisms. In conclusion, lead exposure remains a health problem for both male and female reproductive health. It is important to implement protective measures to avoid the harmful effects of this toxic metal on reproductive health of both men and women.

Acknowledgements

The authors are grateful to researchers, managers, and technicians, who contributed and collaborated in this research. The authors are grateful to researchers Eloisa Esquivel, Gonzalo García Vargas, Ada Sandoval Carrillo, Edna Mendez Hernandez, and Francisco Castellanos Juárez for the assistance in the preparation of this chapter. The authors would also like to recognize the Council of Science and Technology of the State of Durango (COCYTED) for the support to their investigations.

Conflict of interest

The authors declare no conflicts of interest.

Author details

Osmel La Llave León* and José M. Salas Pacheco
Institute of Scientific Research at Juarez University of Durango State, Durango, Mexico

*Address all correspondence to: ollave56@yahoo.es; olallavel@ujed.mx

References

[1] Borja-Aburto VH, Hertz-Picciotto I, Rojas-Lopez M, Farias P, Rios C, Blanco J. Blood lead levels measured prospectively and risk of spontaneous abortion. American Journal of Epidemiology. 1999;**150**:590-597

[2] Irgens A, Krüger K, Skorve AH, Irgens LM. Reproductive outcome in offspring of parents occupationally exposed to lead in Norway. American Journal of Industrial Medicine. 1998;**34**:431-437. DOI: 10.1002/(SICI)1097-0274(199811)34:5<431::AID-AJIM3>3.0.CO;2-T

[3] Cheng L, Zhang B, Huo W, Cao Z, Liu W, Liao J, et al. Fetal exposure to lead during pregnancy and the risk of preterm and early-term deliveries. International Journal of Hygiene and Environmental Health. 2017;**220**:984-989. DOI: 10.1016/j.ijheh.2017.05.006

[4] Rahman A, Kumarathasan P, Gomes J. Infant and mother related outcomes from exposure to metals with endocrine disrupting properties during pregnancy. Science of the Total Environment. 2016;**569-570**:1022-1031. DOI: 10.1016/j.scitotenv.2016.06.134

[5] Huang S, Xia W, Sheng X, Qiu L, Zhang B, Chen T, et al. Maternal lead exposure and premature rupture of membranes: A birth cohort study in China. BMJ Open. 2018;**8**:1-7. DOI: 10.1136/bmjopen-2018-021565

[6] Sowers M, Jannausch M, Scholl T, Li W, Kemp FW, Bogden JD. Blood lead concentrations and pregnancy outcomes. Archives of Environmental Health. 2002;**57**:489-495

[7] Bede-Ojimadu O, Amadi CN, Orisakwe OE. Blood lead levels in women of child-bearing age in sub-Saharan Africa: A systematic review. Frontiers in Public Health. 2018;**6**:1-18. DOI: 10.3389/fpubh.2018.00367

[8] Tyrrell JB, Hafida S, Stemmer P, Adhami A, Leff T. Lead (Pb) exposure promotes diabetes in obese rodents. Journal of Trace Elements in Medicine and Biology. 2017;**39**:221-226. DOI: 10.1016/j.jtemb.2016.10.007

[9] Cantonwine D, Hu H, Sánchez BN, Lamadrid-Figueroa H, Smith D, Ettinger AS, et al. Critical windows of fetal lead exposure. Journal of Occupational and Environmental Medicine. 2010;**52**:1106-1111. DOI: 10.1097/jom.0b013e3181f86fee

[10] Wu HM, Lin-Tan DT, Wang ML, Huang HY, Lee CL, Wang HS, et al. Lead level in seminal plasma may affect semen quality for men without occupational exposure to lead. Reproductive Biology and Endocrinology. 2012;**10**:1. DOI: 10.1186/1477-7827-10-91

[11] Alexander BH, Checkoway H, Van Netten C, Muller CH, Ewers TG, Kaufman JD, et al. Semen quality of men employed at a lead smelter. Occupational and Environmental Medicine. 1996;**53**:411-416. DOI: 10.1136/oem.53.6.411

[12] Debnath B, Ibrahim M, Fatima P. Study of blood lead and semen lead concentration in male infertility. Pulse. 2011;**4**:10-13. DOI: 10.3329/pulse.v4i1.6956

[13] Hosni H, Selim O, Abbas M, Fathy A. Semen quality and reproductive endocrinal function related to blood lead levels in infertile painters. Andrologia. 2013;**45**:120-127. DOI: 10.1111/j.1439-0272.2012.01322.x

[14] Benoff S, Centola GM, Millan C, Napolitano B, Marmar JL, Hurley IR. Increased seminal plasma lead levels adversely affect the fertility potential of sperm in IVF. Human Reproduction. 2003;**18**:374-383. DOI: 10.1093/humrep/deg020

[15] Erfurth EM, Gerhardsson L, Nilsson A, Rylander L, Schütz A, Skerfving S, et al. Effects of lead on the endocrine system in lead smelter workers. Archives of Environmental Health. 2001;**56**:449-455. DOI: 10.1080/00039890109604481

[16] Gulson BLL, Jameson CWW, Mahaffey KRR, Mizon KJJ, Korsch MJJ, Vimpani G. Pregnancy increases mobilization of lead from maternal skeleton. The Journal of Laboratory and Clinical Medicine. 1997;**130**:51-62. DOI: 10.1016/S0022-2143(97)90058-5

[17] Rodamilans M, Osaba MJM, To-Figueras J, Fillat FR, Marques JM, Pérez P, et al. Lead toxicity on endocrine testicular function in an occupationally exposed population. Human & Experimental Toxicology. 1988;**7**:125-128. DOI: 10.1177/096032718800700203

[18] Tuppurainen M, Wägar G, Kurppa K, Sakari W, Fröseth B, Alho J, et al. Thyroid function as assessed by routine laboratory tests of workers with long-term lead exposure. Scandinavian Journal of Work, Environment & Health. 1988;**14**:175-180. DOI: 10.5271/sjweh.1934

[19] Nigg JT, Elmore AL, Natarajan N, Friderici KH, Nikolas MA. Variation in an iron metabolism gene moderates the association between blood lead levels and attention-deficit/hyperactivity disorder in children. Psychological Science. 2016;**27**:257-269. DOI: 10.1177/0956797615618365

[20] Doumouchtsis KK, Doumouchtsis SK, Doumouchtsis EK, Perrea DN. The effect of lead intoxication on endocrine functions. Journal of Endocrinological Investigation. 2009;**32**:175-183. DOI: 10.1007/BF03345710

[21] Haghighi KS, Aminian O, Chavoshi F, Bahaedini LS, Soltani S, Najarkolaei FR. Relationship between blood lead level and male reproductive hormones in male lead exposed workers of a battery factory: A cross-sectional study. International Journal of Reproductive Biomedicine. 2013;**11**:673-676

[22] Al-Omary HL, Alawa ZM, Jaafar I. Environmental lead exposure and male infertility. IOSR Journal of Dental and Medical Sciences. 2016;**15**:49-54. DOI: 10.9790/0853-1509044954

[23] Sokol RZ. Hormonal effects of lead acetate in the male rat: Mechanism of action1. Biology of Reproduction. 1987;**37**:1135-1138. DOI: 10.1095/biolreprod37.5.1135

[24] Ng TP, Goh HH, Ng YL, Ong HY, Ong CN, Chia KS, et al. Male endocrine functions in workers with moderate exposure to lead. British Journal of Industrial Medicine. 1991;**48**:485-491. DOI: 10.1136/oem.48.7.485

[25] Benedetti S, Tagliamonte MC, Catalani S, Primiterra M, Canestrari F, Stefani S De, et al. Differences in blood and semen oxidative status in fertile and infertile men, and their relationship with sperm quality. Reproductive Biomedicine Online 2012;**25**:300-306. DOI: 10.1016/j.rbmo.2012.05.011

[26] Goyer RA, Clarkson T. Toxic effects of metals. In: Klaassen C, editor. Essentials of Toxicology. 6th ed. New York: McGraw-Hill; 2001. p. 811-867

[27] Wiebe JP, Salhanick AI, Myers KI. Life Sciences. Vol. 32. USA: Pergamon Press; 2005. pp. 1997-2005

[28] Batarseh LI, Welsh MJ, Brabec MJ. Effect of lead acetate on sertoli cell lactate production and protein synthesis in vitro. Cell Biology and Toxicology. 1986;**2**:283-292. DOI: 10.1007/BF00122696

[29] Winder C. Reproductive and chromosomal effects of occupational exposure to lead in the male.

Reproductive Toxicology. 1989;**3**:221-233. DOI: 10.1016/0890-6238(89)90016-6

[30] Hsu PC, Liu MY, Hsu CC, Chen LY, Guo YL. Lead exposure causes generation of reactive oxygen species and functional impairment in rat sperm. Toxicology. 1997;**122**:133-143. DOI: 10.1016/S0300-483X(97)00090-5

[31] Siddiqui MK, Srivastava S, Mehrotra PK. Environmental exposure to lead as a risk for prostate cancer. Biomedical and Environmental Sciences. 2002;**15**:298-305

[32] Verze P, Cai T, Lorenzetti S. The role of the prostate in male fertility, health and disease. Nature Reviews. Urology. 2016;**13**:379-386. DOI: 10.1038/nrurol.2016.89

[33] Kelada SN, Shelton E, Kaufmann RB, Khoury MJ. δ-Aminolevulinic acid dehydratase genotype and lead toxicity: A HuGE review. American Journal of Epidemiology. 2001;**154**:1-13. DOI: 10.1126/science.3.53.32

[34] Vigeh M, Saito H, Sawada S. Lead exposure in female workers who are pregnant or of childbearing age. Industrial Health. 2011;**49**:255-261. DOI: 10.2486/indhealth.MS1192

[35] Xu B, Chia SE, Tsakok M, Ong CN. Trace elements in blood and seminal plasma and their relationship to sperm quality. Reproductive Toxicology. 1993;7:613-618

[36] El-Zohairy EA, Youssef AF, Abul-Nasr SM, Fahmy IM, Salem D, Kahil AK, et al. Reproductive hazards of lead exposure among urban Egyptian men. Reproductive Toxicology. 1996;**10**:145-151. DOI: 10.1016/0890-6238(95)02057-8

[37] Robins TG, Bornman MS, Ehrlich RI, Cantrell AC, Pienaar E, Vallabh J, et al. Semen quality and fertility of men employed in a South African lead acid battery plant. American Journal of Industrial Medicine. 1997;**32**:369-376. DOI: 10.1002/(SICI)1097-0274(199710)32:4<369::AID-AJIM8>3.0.CO;2-P

[38] Popovic M, McNeill FE, Chettle DR, Webber CE, Lee CV, Kaye WE. Impact of occupational exposure on lead levels in women. Environmental Health Perspectives. 2005;**113**:478-484. DOI: 10.1289/ehp.7386

[39] La-Llave-León O, Salas Pacheco JM, Estrada Martínez S, Esquivel Rodríguez E, Castellanos Juárez FX, Sandoval Carrillo A, et al. The relationship between blood lead levels and occupational exposure in a pregnant population. BMC Public Health. 2016;**16**:1231. DOI: 10.1186/s12889-016-3902-3

[40] La-Llave-Leon O, Estrada-Martinez S, Salas-Pacheco JM, Pena-Elosegui R, Duarte-Sustaita J, Rangel JLC, et al. Blood Lead levels and risk factors in pregnant women from Durango, Mexico. Archives of Environmental & Occupational Health. 2011;**66**:107-113. DOI: 10.1080/19338244.2010.511311

[41] Kaličanin B, Velimirović D. A study of the possible harmful effects of cosmetic beauty products on human health. Biological Trace Element Research. 2016;**170**:476-484. DOI: 10.1007/s12011-015-0477-2

[42] Nourmoradi H, Foroghi M, Farhadkhani M, Dastjerdi MV. Assessment of lead and cadmium levels in frequently used cosmetic products in Iran. Journal of Environmental and Public Health. 2013;**2013**:2-7. DOI: 10.1155/2013/962727

[43] Al-Saleh I, Al-Enazi S, Shinwari N. Assessment of lead in cosmetic products. Regulatory Toxicology and Pharmacology.

2009;**54**:105-113. DOI: 10.1016/j.yrtph.2009.02.005

[44] Phipps A, Fels H, Burns MS, Gerstenberger SL. Lead poisoning due to geophagia: The consumption of miniature pottery. Open Journal of Pediatrics. 2012;**02**:60-66. DOI: 10.4236/ojped.2012.21010

[45] Thihalolipavan S, Candalla BM, Ehrlich J. Examining pica in NYC pregnant women with elevated blood lead levels. Maternal and Child Health Journal. 2013;**17**:49-55. DOI: 10.1007/s10995-012-0947-5

[46] Bellinger DC. Teratogen update: Lead and pregnancy. Birth Defects Research Part A – Clinical and Molecular Teratology. 2005;**73**:409-420. DOI: 10.1002/bdra.20127

[47] Barry PS, Mossman DB. Lead concentrations in human tissues. British Journal of Industrial Medicine. 1970;**27**:339-351. DOI: 10.1136/oem.27.4.339

[48] Rădulescu A, Lundgren S. A pharmacokinetic model of lead absorption and calcium competitive dynamics. Scientific Reports. 2019;**9**:1-38. DOI: 10.1038/s41598-019-50654-7

[49] Gulson BL, Mizon KJ, Korsch MJ, Palmer JM, Donnelly JB. Mobilization of lead from human bone tissue during pregnancy and lactation - a summary of long-term research. Science of the Total Environment. 2003;**303**:79-104. DOI: 10.1016/S0048-9697(02)00355-8

[50] Gulson B, Taylor A, Eisman J. Bone remodeling during pregnancy and post-partum assessed by metal lead levels and isotopic concentrations. Bone. 2016;**89**:40-51. DOI: 10.1016/j.bone.2016.05.005

[51] Ying WY, Xu SK, Li H, Yan MH. The effects of lead exposure on placental NF-κB expression and the consequences for gestation. Reproductive Toxicology. 2009;**27**:190-195. DOI: 10.1016/j.reprotox.2008.12.006

[52] Gundacker C, Hengstschläger M. The role of the placenta in fetal exposure to heavy metals. Wiener Medizinische Wochenschrift. 2012;**162**:201-206. DOI: 10.1007/s10354-012-0074-3

[53] Hertz-Picciotto I. The evidence that lead increases the risk for spontaneous abortion. American Journal of Industrial Medicine. 2000;**38**:300-309. DOI: 10.1002/1097-0274(200009)38:3<300::AID-AJIM9>3.0.CO;2-C

[54] Vigeh M, Yokoyama K, Seyedaghamiri Z, Shinohara A, Matsukawa T, Chiba M, et al. Blood lead at currently acceptable levels may cause preterm labour. Occupational and Environmental Medicine. 2011;**68**:231-234. DOI: 10.1136/oem.2009.050419

[55] Taylor CM, Golding J, Emond AM. Adverse effects of maternal lead levels on birth outcomes in the ALSPAC study: A prospective birth cohort study. BJOG: An International Journal of Obstetrics and Gynaecology. 2015;**122**:322-328. DOI: 10.1111/1471-0528.12756

[56] Soomro MH, Baiz N, Huel G, Yazbeck C, Botton J, Heude B, et al. Exposure to heavy metals during pregnancy related to gestational diabetes mellitus in diabetes-free mothers. Science of the Total Environment. 2019;**656**:870-876. DOI: 10.1016/j.scitotenv.2018.11.422

[57] Wells EM, Navas-Acien A, Herbstman JB, Apelberg BJ, Silbergeld EK, Caldwell KL, et al. Low-level lead exposure and elevations in blood pressure during pregnancy. Environmental Health Perspectives. 2011;**119**:664-669. DOI: 10.1289/ehp.1002666

[58] Vigeh M, Yokoyama K, Mazaheri M, Beheshti S, Ghazizadeh S, Sakai T, et al. Relationship between increased

blood lead and pregnancy hypertension in women without occupational lead exposure in Tehran, Iran. Archives of Environmental Health. 2004;**59**:70-75. DOI: 10.3200/AEOH.59.2.70-75

[59] Yoon JH, Ahn YS. The association between blood lead level and clinical mental disorders in fifty thousand lead-exposed male workers. Journal of Affective Disorders. 2016;**190**:41-46. DOI: 10.1016/j.jad.2015.09.030

[60] Bayat F, Akbari SAA, Dabirioskoei A, Nasiri M, Mellati A. The relationship between blood Lead level and preeclampsia. Electronic Physician. 2016;**8**:3450-3455. DOI: 10.19082/3450

[61] Dawson EB, Evans DR, Kelly R, Van Hook JW. Blood cell lead, calcium, and magnesium levels associated with pregnancy-induced hypertension and preeclampsia. Biological Trace Element Research. 2000;**74**:107-116. DOI: 10.1385/BTER:74:2:107

[62] Disha SS, Goyal M, Kumar PK, Ghosh R, Sharma P. Association of raised blood lead levels in pregnant women with preeclampsia: A study at tertiary centre. Taiwanese Journal of Obstetrics & Gynecology. 2019;**58**: 60-63. DOI: 10.1016/j.tjog.2018.11.011

[63] Ikechukwu IC, Ojareva OIA, Ibhagbemien AJ, Okhoaretor OF, Oluwatomi OB, Akhalufo OS, et al. Blood lead, calcium, and phosphorus in women with preeclampsia in Edo State, Nigeria. Archives of Environmental and Occupational Health. 2012;**67**:163-169. DOI: 10.1080/19338244.2011.619212

[64] Vigeh M, Yokoyama K, Ramezanzadeh F, Dahaghin M, Sakai T, Morita Y, et al. Lead and other trace metals in preeclampsia: A case-control study in Tehran, Iran. Environmental Research. 2006;**100**:268-275. DOI: 10.1016/j.envres.2005.05.005

[65] Vigeh M, Yokoyama K, Shinohara A, Afshinrokh M, Yunesian M. Early pregnancy blood lead levels and the risk of premature rupture of the membranes. Reproductive Toxicology. 2010;**30**:477-480. DOI: 10.1016/j.reprotox.2010.05.007

[66] Huang S, Xia W, Sheng X, Qiu L, Zhang B, Chen T, et al. Maternal lead exposure and premature rupture of membranes: A birth cohort study in China. BMJ Open. 2018;**8**:e021565. DOI: 10.1136/bmjopen-2018-021565

[67] Srivastava S, Mehrotra PK, Srivastava SP, Tandon I, Siddiqui MKJ. Blood lead and zinc in pregnant women and their offspring in intrauterine growth retardation cases. Journal of Analytical Toxicology. 2001;**25**:461-465. DOI: 10.1093/jat/25.6.461

[68] Zhu M, Fitzgerald EF, Gelberg KH, Lin S, Druschel CM. Maternal low-level lead exposure and fetal growth. Environmental Health Perspectives. 2010;**118**:1471-1475. DOI: 10.1289/ehp.0901561

[69] Zhang B, Xia W, Li Y, Bassig BA, Zhou A, Wang Y, et al. Prenatal exposure to lead in relation to risk of preterm low birth weight: A matched case-control study in China. Reproductive Toxicology. 2015;**57**:190-195. DOI: 10.1016/j.reprotox.2015.06.051

[70] Murphy MJ, Graziano JH, Popovac D, Kline JK, Mehmeti A, Factor-Litvak P, et al. Past pregnancy outcomes among women living in the vicinity of a lead smelter in Kosovo, Yugoslavia. American Journal of Public Health. 1990;**80**:33-35. DOI: 10.2105/AJPH.80.1.33

[71] Tabacova S, Balabaeva L. Environmental pollutants in relation to complications of pregnancy. Environmental Health Perspectives. 1993;**101**:27-31. DOI: 10.2307/3431372

[72] Hertz-Picciotto I, Schramm M, Watt-Morse M, Chantala K, Anderson J, Osterloh J. Patterns and determinants of blood lead during pregnancy. American Journal of Epidemiology. 2000;**152**:829-837. DOI: 10.1093/aje/152.9.829

[73] Silbergeld EK, Schwartz J, Mahaffey K. Lead and osteoporosis: Mobilization of lead from bone in postmenopausal women. Environmental Research. 1988;**47**:79-94. DOI: 10.1016/S0013-9351(88)80023-9

[74] Potula V, Kaye W. Is lead exposure a risk factor for bone loss? Journal of Women's Health. 2005;**14**:461-464. DOI: 10.1089/jwh.2005.14.461

[75] Sanín LH, González-Cossío T, Romieu I, Peterson KE, Ruíz S, Palazuelos E, et al. Effect of maternal lead burden on infant weight and weight gain at one month of age among breastfed infants. Pediatrics. 2001;**107**:1016-1023. DOI: 10.1542/peds.107.5.1016

[76] González-Cossío T, Peterson KE, Sanín LH, Fishbein E, Palazuelos E, Aro A, et al. Decrease in birth weight in relation to maternal bone-lead burden. Pediatrics. 1997;**100**:856-862. DOI: 10.1542/peds.100.5.856

[77] Oguri T, Ebara T, Nakayama SF, Sugiura-Ogasawara M, Kamijima M, Saito H, et al. Association between maternal blood cadmium and lead concentrations and gestational diabetes mellitus in the Japan Environment and Children's Study. International Archives of Occupational and Environmental Health. 2019;**92**:209-217. DOI: 10.1007/s00420-018-1367-7

[78] Vaziri ND, Gonick HC. Cardiovascular effects of lead exposure. The Indian Journal of Medical Research. 2008;**128**:426-435

[79] Schwartz J. The relationship between blood lead and blood pressure in the NHANES II survey. Environmental Health Perspectives. 1988;**78**:15-22. DOI: 10.1289/ehp.887815

[80] Navas-Acien A, Guallar E, Silbergeld EK, Rothenberg SJ. Lead exposure and cardiovascular disease - a systematic review. Environmental Health Perspectives. 2007;**115**:472-482. DOI: 10.1289/ehp.9785

[81] Han L, Wang X, Han R, Xu M, Zhao Y, Gao Q, et al. Association between blood lead level and blood pressure: An occupational population-based study in Jiangsu province, China. PLoS One. 2018;**13**:1-10. DOI: 10.1371/journal.pone.0200289

[82] Rothenberg SJ, Kondrasho V, Manalo M, Jiang J, Cuellar R, Garcia M, et al. Increases in hypertension and blood pressure during pregnancy with increased bone lead levels. American Journal of Epidemiology. 2002;**156**: 1079-1087. DOI: 10.1093/aje/kwf163

[83] Magri J, Sammut M, Savona-Ventura C. Lead and other metals in gestational hypertension. International Journal of Gynecology & Obstetrics. 2003;**83**:29-36. DOI: 10.1016/S0020-7292(03)00212-1

[84] Ugwuja E, Ejikeme B, Obuna J. Impacts of elevated prenatal blood lead on trace element status and pregnancy outcomes in occupationally non-exposed women. International Journal of Occupational Environmental Medicine. 2011;**2**:143-156

[85] Hutcheon JA, Lisonkova S, Joseph KS. Epidemiology of pre-eclampsia and the other hypertensive disorders of pregnancy. Best Practice & Research: Clinical Obstetrics & Gynaecology. 2011;**25**:391-403. DOI: 10.1016/j.bpobgyn.2011.01.006

[86] Vaziri ND, Khan M. Interplay of reactive oxygen species and nitric oxide in the pathogenesis of experimental lead-induced hypertension. Clinical

and Experimental Pharmacology & Physiology. 2007;**34**:920-925. DOI: 10.1111/j.1440-1681.2007.04644.x

[87] Vaziri ND. Mechanisms of lead-induced hypertension and cardiovascular disease. American Journal of Physiology. Heart and Circulatory Physiology. 2008;**295**:H454-H465. DOI: 10.1152/ajpheart.00158.2008

[88] Gonick HC, Ding Y, Bondy SC, Ni Z, Vaziri ND. Lead-induced hypertension: Interplay of nitric oxide and reactive oxygen species. Hypertension. 1997;**30**:1487-1492. DOI: 10.1161/01.HYP.30.6.1487

[89] Carmignani M, Volpe AR, Boscolo P, Qiao N, Di Gioacchino M, Grilli A, et al. Catcholamine and nitric oxide systems as targets of chronic lead exposure in inducing selective functional impairment. Life Sciences. 2000;**68**:401-415. DOI: 10.1016/S0024-3205(00)00954-1

[90] Moreau T, Hannaert P, Orssaud G, Huel G, Garay RP, Claude JR, et al. Influence of membrane sodium transport upon the relation between blood lead and blood pressure in a general male population. Environmental Health Perspectives. 1988;**78**:47-51. DOI: 10.1289/ehp.887847

[91] Joo H, Choi JH, Burm E, Park H, Hong YC, Kim Y, et al. Gender difference in the effects of lead exposure at different time windows on neurobehavioral development in 5-year-old children. Science of the Total Environment. 2018;**615**:1086-1092. DOI: 10.1016/j.scitotenv.2017.10.007

[92] Vahter M, Åkesson A, Lidén C, Ceccatelli S, Berglund M. Gender differences in the disposition and toxicity of metals. Environmental Research. 2007;**104**:85-95. DOI: 10.1016/j.envres.2006.08.003

[93] Oishi H, Nomiyama H, Nomiyama K, Tomokuni K. Comparison between males and females with respect to the porphyrin metabolic disorders found in workers occupationally exposed to lead. International Archives of Occupational and Environmental Health. 1996;**68**:298-304. DOI: 10.1007/BF00409414

[94] Chang SH, Cheng BH, Lee SL, Chuang HY, Yang CY, Sung FC, et al. Low blood lead concentration in association with infertility in women. Environmental Research. 2006;**101**:380-386. DOI: 10.1016/j.envres.2005.10.004

[95] Keitt SK, Fagan TF, Marts SA. Understanding sex, differences in environmental health: A thought leaders' roundtable. Environmental Health Perspectives. 2004;**112**:604-609. DOI: 10.1289/ehp.6714

[96] Llop S, Lopez-Espinosa MJ, Rebagliato M, Ballester F. Gender differences in the neurotoxicity of metals in children. Toxicology. 2013;**311**:3-12. DOI: 10.1016/j.tox.2013.04.015

[97] Kumar S. Occupational, environmental and lifestyle factors associated with spontaneous abortion. Reproductive Sciences. 2011;**18**:915-930. DOI: 10.1177/1933719111413298

[98] Main KM, Skakkebæk NE, Virtanen HE, Toppari J. Genital anomalies in boys and the environment. Best Practice & Research. Clinical Endocrinology & Metabolism. 2010;**24**:279-289. DOI: 10.1016/j.beem.2009.10.003

[99] Escobar-Morreale HF. Polycystic ovary syndrome: Definition, aetiology, diagnosis and treatment. Nature Reviews. Endocrinology. 2018;**14**:270-284. DOI: 10.1038/nrendo.2018.24

[100] Cole TJ, Ahmed ML, Preece MA, Hindmarsh P, Dunger DB. The relationship between insulin-like growth Factor 1, sex steroids and timing of the pubertal growth spurt. Clinical

Endocrinology. 2015;**82**:862-869. DOI: 10.1111/cen.12682

[101] Sergeyev O, Burns JS, Williams PL, Korrick SA, Lee MM, Revich B, et al. The association of peripubertal serum concentrations of organochlorine chemicals and blood lead with growth and pubertal development in a longitudinal cohort of boys: A review of published results from the Russian Children's Study. Reviews on Environmental Health. 2017;**32**:83-92. DOI: 10.1515/reveh-2016-0052

[102] Yao Q, Zhou G, Xu M, Dai J, Qian Z, Cai Z, et al. Blood metal levels and serum testosterone concentrations in male and female children and adolescents: NHANES 2011-2012. PLoS One. 2019;**14**:1-14. DOI: 10.1371/journal.pone.0224892

[103] Liu Y, Téllez-Rojo MM, Sánchez BN, Zhang Z, Afeiche MC, Mercado-García A, et al. Early lead exposure and pubertal development in a Mexico City population. Environment International. 2019;**125**:445-451. DOI: 10.1016/j.envint.2019.02.021

[104] Wani AL, Ara A, Usmani JA. Lead toxicity: A review. Interdisciplinary Toxicology. 2015;**8**:55-64. DOI: 10.1515/intox-2015-0009

[105] Markowitz M. Lead poisoning: A disease for the next millennium. Current Problems in Pediatrics. 2000;**30**:62-70. DOI: 10.1067/mps.2000.104053

[106] Obeng-Gyasi E. Sources of lead exposure in various countries. Reviews on Environmental Health. 2019;**34**: 25-34. DOI: 10.1515/reveh-2018-0037

[107] La-Llave-León O, Méndez-Hernández EM, Castellanos-Juárez FX, Esquivel-Rodríguez E, Vázquez-Alaniz F, Sandoval-Carrillo A, et al. Association between blood lead levels and delta-aminolevulinic acid dehydratase in pregnant women. International Journal of Environmental Research and Public Health. 2017;**14**:1-10. DOI: 10.3390/ijerph14040432

Chapter 2

Epigenetics and Lead Neurotoxicity

Yi Xu, Tian Wang and Jie Zhang

Abstract

Lead exposure continues to threaten human health in a worldwide perspective. Among the multiple target organs affected by lead, central nervous system (CNS) pervades in the adverse consequences by chronic lead exposure, leading to a variety of neurotoxic manifestations and neurological disorders. The epigenetic machinery plays a vital role in the control of key neural functions, particularly neuronal development. Faulty epigenetic gene regulation can have marked deleterious effect on the developing brain that can last for an entire lifespan. Mounting evidence suggests that lead exposure can pose detrimental effect on CNS through these epigenetic mechanisms. And this chapter reviews the current understandings of concrete epigenetic forms, exemplified by DNA methylation, histone modification, and ncRNAs, responding to lead exposure and moderating the consequent neurotoxicity. In addition, Alzheimer's disease (AD) is presented as a typical instance to explain how environmental lead exposure results in the occurrence of AD in an "early exposure, late onset" fashion. A future perspective, highlighting additional forms of epigenetic elements as well as interactive actions among different molecules, was also proposed. In summary, epigenetics was substantially implicated in regulating lead neurotoxicity.

Keywords: epigenetics, lead neurotoxicity, DNA methylation, histone modification, ncRNA, Alzheimer's disease

1. Introduction

Lead (Pb) is a ubiquitous and persistent neurotoxicant that continues to threaten human health in a global perspective [1]. Although lead has been removed from paints and gasoline, it remains a serious concern as it can still be found in a variety of daily products, including toys, batteries, food, and water [2]. Although Pb poisoning is a preventable disease, thousands of new cases in the United States were reported each year, and about 500,000 children under 5 years old have blood lead levels (BLLs) greater than a threshold level of 5 g/dl, according to Centers for Disease Control and Prevention (CDC) reports [3–7]. Elevated BLLs have become the first noninfection condition to be notifiable at the national level [7]. Lead can cause a series of adverse human consequences at a very low level exposure [8]. Therefore, CDC continually decreased the safe threshold of BLLs, and the current "safe" levels of exposure to lead are 5 mg/dl for children, but still there have been studies to identify cognitive impairments below that dosage, implying that "no level of lead exposure is safe" [6, 9].

IntechOpen

Lead pervades many organs and systems in the human body, but the prime target of lead toxicity is CNS, both in adults and in children [10], resulting in the so-called "neurotoxicity." The developing brain is particularly susceptible to lead neurotoxicity, as demonstrated by several epidemiological and experimental studies [2, 3]. Due to the fact that lead can freely cross blood brain barrier, lead neurotoxicity can also be manifested in adults, with a larger exposure dosage. Particularly, it should be noteworthy that early life exposure to lead can produce persistent alterations in the brain structure of adults, causing lasting impairment of brain function and behavior [3, 11]. Adverse neurotoxic effects caused by lead include intellectual and behavioral deficits in children; deficits in fine motor function and coordination; and deficits in lower performance on intelligence tests [10]. Higher level of lead can cause a wide spectrum of neurological disorders, such as convulsions and coma, including multiple instances of neurodegenerative disorders, such as AD and Parkinson's disease [12–15]. Thus, there is a critical need to understand the mechanisms of lead neurotoxicity.

Among the cellular and molecular mechanisms suggested to underlie lead neurotoxicity, amounting evidence underscored roles of epigenetic molecules. This fast-moving filed of epigenetics has opened a novel avenue of research for understanding how environmentally toxic signals like lead exposure could be readily sensed by organisms and then relayed to reprogram the expression of key functional genes, consequently giving rise to neurotoxic manifestations [3, 16–19]. This chapter is aimed to discuss the advances of epigenetic alterations in response to lead-induced neural deficits, focusing on the concrete epigenetic species and their responsive details. We will also present a synoptic view of epigenetic implications in etiology of AD caused by long-term lead exposure and bring out the possible future perspectives of the related research topics.

2. Molecular mechanism of lead-induced neurotoxicity

Since lead was found to mediate severe neurological impairment toward both children and adults, myriad studies were appreciated to decipher the cellular and molecular alterations underlying this neurotoxic incident. Several routes of action have been most commonly proposed, such as oxidative stress, disruption of blood brain barrier, decreased cellular energy metabolism, deregulation of calcium signaling, and abnormal neural transmission [3, 20]. In terms of relevance to neuronal development and synaptic transmission, postsynaptic mechanisms represented by N-methyl-D-aspartate receptor (NMDAR), presynaptic mechanisms, and brain-derived neurotrophic factor (BDNF) signaling were shown to be involved in lead neurotoxicity [8, 21].

NMDR plays an essential role in hippocampus-mediated learning and memory, and its dysfunction is associated with spatial learning abnormalities, as well as dendritic atrophies [22]. Lead regularly disrupted NMDAR function by acting as a potent antagonist. Apart from it, lead exposure also disrupts normal NMDAR ontogeny, such as reducing NR2A content, altering expression of NR1 splice variants [23, 24].

Chronic lead exposure also results in impaired neurotransmission. A previous finding showed that chronic lead exposure reduced Ca^{2+}-dependent glutamate and γ-aminobutyric acid (GABA) release in the rat hippocampus [25, 26]. And in cultured hippocampal neurons, lead exposure was found to impair excitatory postsynaptic currents (EPSCs) and inhibitory postsynaptic currents (IPSCs) [27]. Our lab recently published a finding that chronic lead exposure can inhibit the release of neurotransmitters by interfering with its vesicle pool recycling, and the main protein impacted is synapsin 1, which expression and phosphorylation was prone to lead invasion [28].

An emerging theme involved in lead neurotoxicity is the disruption of brain-derived neurotrophic factor (BDNF) expression. BDNF is a trans-synaptic signaling molecule that is released from both dendrites and axons [29]. In response to lead exposure, BDNF levels in cell cultures were downregulated, and the exogenous addition of BDNF can rescue the deleterious effect of Pb [30]. As a regulator of Ca^{2+} signaling and homeostasis, BDNF perturbation in turn led to the disrupted Ca^{2+}-dependent pathways, which compromise severe neural representations caused by lead exposure [31]. As an alternative consequence, lead impaired the hippocampal dendritic spines, reduced their density, and changed their morphology [32].

These neuronal and molecular processes might reflect variable aspects of lead-induced neurotoxicity. Compared to it, roles of global regulators, such as the emerging epigenetic regulators, are not sufficiently understood. However, given the conformity with outstanding characteristics of lead neurotoxicity, like "early exposure, persisted effect," epigenetics was long hypothesized to be implicated in the etiology of lead-induced psychological disorders [3]. This was supported by further identification of lead exposure as a risk factor of Alzheimer's disease and schizophrenia. Our next section will focus on epigenetic determinants involved in lead-led neurological damages and diseases [33, 34].

3. Epigenetic mechanisms

Epigenetics is defined as the heritable changes in gene expression that are not related to alterations in the genetic code [3]. Epigenetic regulation is found to modify the conformational state of chromatin and the accessibility of specific gene promoters to the transcriptional machinery [14]. There are three main epigenetic mechanisms broadly studied: DNA methylation, posttranslational modifications of histones, and noncoding RNA (ncRNA) [35–37]. The basic modes of action of three epigenetic forms were shown in **Figure 1**. DNA methylation is the most-studied epigenetic mechanism, which involves primarily cytosine methylation of Cytosine Guanine dinucleotides (CpG) via DNA methyltransferases (DNMTs). CpG methylation is often linked with transcriptional inhibition as it interferes with the normal binding and activity of transcription binding proteins [38]. The exception to this is CpG islands, which are CpG-rich sequences that are densely populated with unmethylated CpGs [2]. CpG islands offer the possibility of being differentially regulated by the environmental signals, which are a prime site to study the influence of lead on epigenetic determinants of ensuing neurotoxic phenotypes [39].

Gene expression is also regulated by histone modifications [36]. Histones are alkaline proteins that wrapped around DNA in nucleosomes and moderated gene transcription by modulating chromatin compaction and accessibility [40]. The terminal tails of histone can undergo covalent posttranslational modifications (PTMs), which in turn alter their interaction with DNA. Diverse modification forms were discovered and studied, as well as their influence on transcription of objective genes, including acetylation, phosphorylation, methylation, and ubiquitination [2]. The complex forms of histone modifications could coexist in regulating a common gene or genome, establishing an intricate and complex regulatory network called "histone code" [41, 42]. The proposal of this definition opens an avenue to show the potential accuracy and delicacy of gene expression regulation, via the action of histone modifications, which are operated by corresponding enzymes, such as histone acetyltransferases, histone deacetylases, histone methylases, and histone demethylases. Among them, histone deacetylases have been widely studied and identified as key molecular targets for pharmaceutical interventions [43, 44].

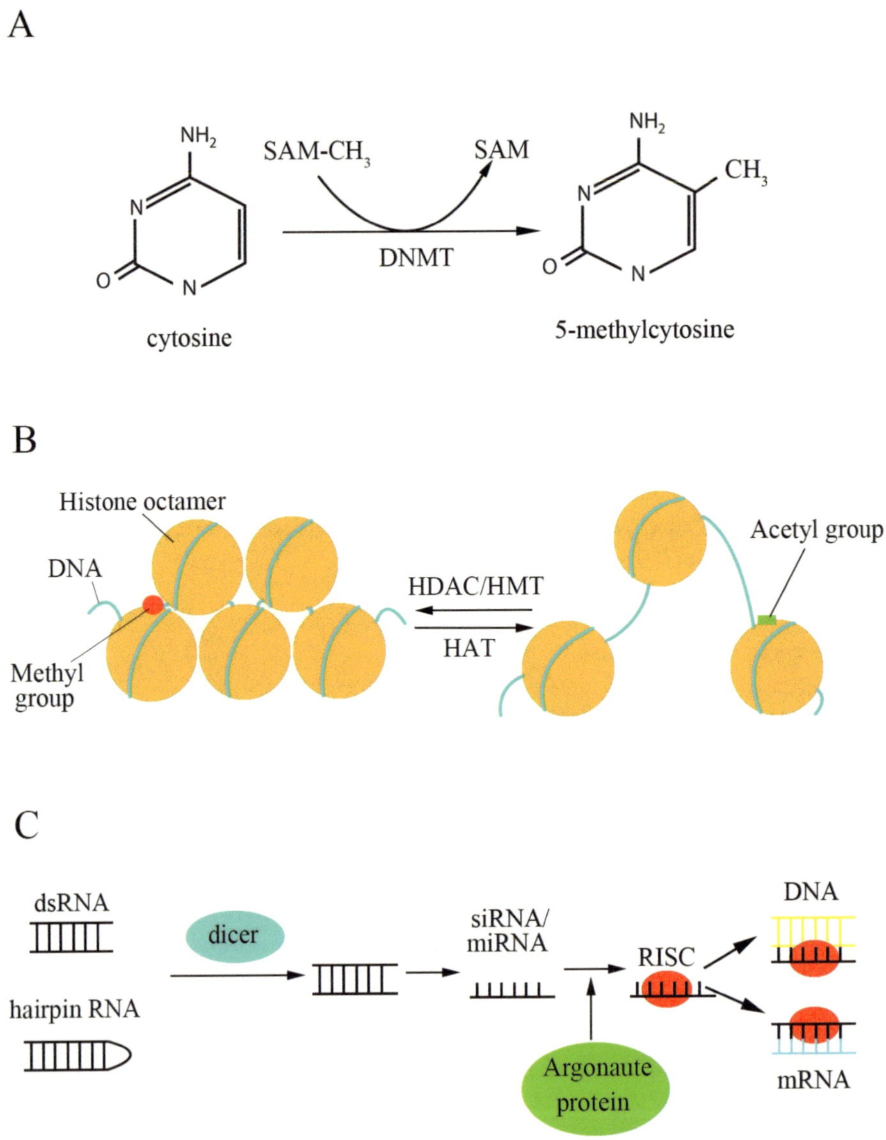

Figure 1.
Types of epigenetic modifications. (A) CpG methylation; (B) histone modifications represented by acetylation and methylation; (C) biogenesis and inhibitory action of microRNA (miRNA).

Another layer of epigenetic regulation involves noncoding RNAs (ncRNAs), defined as functional RNA molecules that are not translated into proteins [3]. The category of ncRNA with relevance with epigenetic regulation is composed of microRNA (miRNA), lncRNA, piRNA, and snoRNA, which regulate gene expression at both transcriptional and posttranscriptional levels [37]. miRNA received the most attention among them, and details of miRNA types involved in various epigenetic regulation and neuronal processes were studied comprehensively. For instance, a recent publication revealed that a series of miRNAs can respond to and mediate spinal muscular atrophy pathogenesis [45]; miR-137 participated in the modulation of neuronal maturation by targeting an ubiquitin ligase mind bomb-1 [46].

The different forms of epigenetic regulation are often functionally interlaced, showing an interactive relationship. DNA methylation was commonly accompanied by a specific form of histone methylation, establishing a so-called "hand-in-hand" assembly [47]. Histone deacetylase can be recruited by DNA methylation or its functional partner, and these molecules can cooperate with each other to determine (most commonly inhibit) the expressional status of the objective gene [48]. DNA methylations and histone modifications are both involved in establishing patterns of gene expression, and they may play distinct roles in inducing persistent or intermittent gene expression changes [14].

Epigenetic mechanisms are implicated in neuronal development, maintenance of cell identity, and aging process [49, 50]. Meanwhile, epigenetic perturbations that lead to chromatin remodeling are in association with a number of neuropsychiatric and neurodegenerative disorders [14, 51], and they are particularly relevant in response to environmental toxicant exposure early in life [38]. The impact of epigenetic determinants can be long lasting, or even transgenerational, that is, the epigenetic traits and gene expression patterns can be sometimes inherited by next generations, which are not previously exposed to the causative agents [52].

4. Lead-induced epigenetic alterations in CNS

Considering that epigenetic factors are associated with CNS functioning and meanwhile susceptible to environmental exposure, we next discuss the epigenetic outcome brought by lead exposure and the roles of these changes in the etiology and development of lead neurotoxicity.

The relations of lead exposure with neural epigenetic alterations have long been established. Developmental lead exposure results in a variety of epigenetic changes, characterized by DNA methylation alterations, which can impact gene expression patterns and affect nervous system development [5]. Lead can promptly affect the dynamism of epigenetic determinants and promote the rapid turnover of DNA methylation [4]. Consistent with these findings and statements, lead is known to be an epigenetic modifier [53]. The following are advances of three main epigenetic systems associated with lead neurotoxicity:

4.1 DNA methylation

Lead exposure can result in the global changes of DNA methylation profiles in CNS, and the detailed orientation of methylome changes varied depending on the studied genetic microenvironment. Singh et al. stated that developmental lead exposure disrupted the hippocampal methylome, and the effect is dependent on the gender, timing, and level of exposure [5]. In particular, the global methylome alterations did not reflect the methylation status of the specific genes. That identified low-level lead exposure as a causative agent of gene-specific DNA methylation patterns in brain [54]. Senut et al. found that lead exposure disrupts global DNA methylation in human embryonic stem cells and alters their neuronal differentiation [4]. An epidemiological study investigated 105 children participants from birth to 78 months and tested peripheral blood DNA to quantify CpG methylation at Differentially Methylated Regions (DMRs) of 22 human imprinted genes. This study provided evidence that early childhood lead exposure resulted in gene-specific DNA methylation differences in the DMRs of PEG3, PLAGL1/HYMAI, and IGF2/H19 [51]. In 2009, Pilsner et al. published the first human study to reveal that maternal bone lead levels were associated with changes in DNA methylation levels in the umbilical cord blood leukocytes of the offspring [55]. In animal studies, DNA methylome-related genes, including DNA methyltransferases, methyl-cytosine-phosphate-guanine (Me-CpG) binding protein-2 (MeCP2), and methionine synthase, were recognized as the potential targets of lead exposure [56, 57]. Sanchez-Martin et al. reported that lead exposure resulted in hypermethylation of three differentially methylated regions in the hippocampus of females, but not males [11]. Overall, lead is a strong environmental force to globally reshape DNA methylation landscape in brain, which is a susceptible organ for epigenetic regulations.

Apart from methylome, the gene-specific alterations of methylation may reflect the detailed influence of lead exposure in CNS. Zawia et al. examined the activity of DNA methyltransferase in the tissues of 23-year-old primates exposed to lead as infants. As a consequence, they found that activity of this methylation enzyme was selective for cytosine nucleotides in a CpG sequence and specific to ones that base-paired to methylated CpG sequence on the other DNA strand [25]. Some genes triggered during memory formation and synaptic plasticity, such as BDNF, showed marked changes in promoter methylation when DNMT activity is suppressed in mice hippocampus, indicating that BDNF can be potentially modulated by specified DNA methylation status [58]. Wu et al. investigated the association between prenatal maternal lead exposure and epigenome-wide DNA methylation. Among female infants, one CpG (cg24637308) showed a strong negative association with lead levels, and this CpG site was thought to be highly expressed in human brain [59]. Our previous study also measured CpG methylation levels in specific CpG-rich promoter regions of DAT1 and DRD4, two dopaminergic-related genes, in the children with higher blood lead levels. According to it, a specific CpG site located upstream of DRD4-coding region was found to be hypermethylated due to lead exposure, and this changes were negatively correlated with the expression levels of DRD4 [39]. The relevant literature pertaining to associations of lead neurotoxicity and DNA methylation was summarized and shown in **Table 1**.

A number of reports suggested that alteration in DNA methylation was largely gender- and tissue-dependent [73]. Early life lead exposure of 3 and 30 ppm led to gender-specific DNA hypermethylation at Rn45a and Sfi1 genes in the hippocampus of female mice only [11]. Another study also stated that maternal lead exposure caused gender-specific epigenetic outcomes for varying degree of vulnerability later in life [74]. These gender differences might be related to the action of sex hormone and the structural discrepancies of body structure resulting in the variance of lead metabolic routes.

CpG methylation was reprogrammed through the action of DNA methylases. Amounting evidence suggested that this pathway was utilized by lead to bring adverse neurological outcome. In a 23-year-old primate with early exposure of lead, protein levels of DNMTs and MeCP2 were significantly decreased. And this attenuation was consistent with hypomethylating effects at multiple genetic loci [14]. Another report found a decreased DNA methyltransferase activity in mouse cortical neuronal cells exposed to lead for 24 h and determined later in life [16]. In a mouse study with developmental lead exposure, DNMT3a in male mice displayed increased expression with 150 and 750 ppm of Pb, while female mice had decreased expression of DNMT3a in response to 150 and 375 ppm of Pb. Moreover, developmental lead exposure can also affect the expression of DNMT1 and MeCP2 in murine hippocampus, which gender and exposure periods were critical contributing factors [57]. Therefore, both expression levels and enzymatic activities of methylation-modifying enzymes can be modulated by lead exposure in CNS.

Except from a conventional methylation form, 5-hydroxymethylated cytosine (5hmC), a new modification and mostly implicated in promoting gene expression, was recently known to be altered by lead exposure in CNS. 5hmCs was extremely abundant in rodent brains, and they are closely associated with critical neurodevelopmental processes such as neuronal differentiation and synaptic function [75]. In light of it, Sen et al. reported that prenatal exposure to lead can alter the hydroxymethylation profile of 5hmC-riched clusters of imprinted genes, which resulted in an

Animal (Age)	Exposure duration	Epigenetic mechanisms	Pathophysiological outcomes (possible)	References
Human embryonic stem cell	A: day 1 B: day 5 C: days 0–19 D: days 11–19	DNA methylation status of genes crucial to brain development	Exposure to Pb subtly alters the neuronal differentiation of exposed hESCs	[4]
Newborns (prenatal)	Prenatal	DNA methylation at LINE-1 repetitive elements; UCB LINE-1 methylation	Epigenetic alterations have detrimental effects on the developing brain, neurological development, and disease	[38]
Newborns	From birth to 78 months	DMR methylation for PEG3 (A), IGF2/H19 (B), and HYMA/PLAGL1 (C)	Sex-dependent and gene-specific DNA methylation	[51]
Mice (10 months)	2 weeks prior mating and during gestation and lactation until PND21	Average brain methylation: IAP 110:80%	Neurodegeneration, narcolepsy	[60]
Mice (PND 20 and 700)	Gestational D13 until PND20	DNA hypermethylation	Gene repression in old age	[61]
Rats (PND55)	10 days prior breeding till weaning of 55D	↓DNMT1 with postnatal 150ppm Pb ↓DNMT1 with postnatal 375ppm Pb ↑DNMT1 with postnatal 750ppm Pb ↓DNMT1 will all perinatal Pb	Defects in neuronal maturation, synaptic plasticity, learning, memory, cognition, and behavior	[57]
Mice (PND20 and PND700)	Gestational D13 until PND20	501 downregulated genes and 647 upregulated genes	Affecting immune responses, metal binding, metabolism, transcription, and transduction	[56]
Mice (PND1 to PND20)	PND1 to PND21	↑Dlx1 methylation ↓Gene expression of Dlx1/2/5/6 ↑Gene expression of Tubb3	Hyperactivity, weight loss, abnormal behavior	[62]
Monkeys (23 years)	Birth until 400D	↓DNMT1 ↑APPmRNA ↑Aβ1-40mRNA ↑Aβ1-42mRNA	Alzheimer's disease	[16]
Mice (2 months)	2 months prior mating and during gestation and lactation until PND2	Hypermethylation in: Rn4.5s loci in chromosome2 Sfi1 loci in chromosome11 (Rn45s loci inchromosome17)	DNA methylation (sex and tissue specific)	[11]

Animal (Age)	Exposure duration	Epigenetic mechanisms	Pathophysiological outcomes (possible)	References
Mice (PND0 and PND6)	2 months prior breeding and throughout lactation	↓H3K9/14Ac:b/w PND0 & PND6 (HPC) ↓ H3K9Me3:b/w PND0 & PND6 ↓H3K9/14Ac ↓H3K9Me3 ↓H3K9/14Ac: 50% at PND0 with 100ppm Pb ↑H3K9/14Ac: 60% at PND0 with 100ppm Pb	Weight loss, abnormal brain development and cognitive function (sex, age, and brain region specific)	[63]
Rats	Perinatal to 60D	Acetylated H3 in hippocampus ↑p300 (HAT) mRNA ↑HDAC1 mRNA	Hyperactivity, behavioral disorder, neurological disorder, ADHD (dose specific)	[64]
Monkeys (3–6,12,23 years)	Birth until 400D	Altered gene expression (22) ↑APP mRNA and protein ↓Dnmt3a and Dnmt1 at 23years ↓MeCP2 at 23years ↑H3K9ac, H4K8ac, H4K12ac ↑H3K4me2	Neurodegeneration in old age, up and downregulation of genes	[14]
Mice (Weeks 20)	E1 to E10	Hypomethylated Chd7 gene ↑Chd7 gene expression (4.7 folds) Altered histone methylation	Autism-like behavior, multiple behavioral abnormalities	[65]
Mice (PND20, 180,270, 540, and 700)	PND1 to PND20	↓DNMT1 protein ↑DNMT3a mRNA at PND20 ↓MeCP2 at PND20 and 270 ↑MAT2A at PND270, 540 and 700 ↓H3K9Ac protein at PND700 ↓H3K4Me2 protein at PND20	Alzheimer disease	[15]
Transgenic mice (15) (PND20, 30, 40, and 60)	PND1 to PND20	Hyperphosphorylation (internal and external brain capsule) ↑ miR34c expression b/w PND20 and 50 ↑ tau mRNA at PND20 ↑CDK5 mRNA at PND40 ↑ Total tau protein at PND20 and 40 ↑CDK5 protein at PND40 and 60 ↑ Phosphorylated tau Ser396 protein at PND20 and 30	Alzheimer's disease, tauopathies	[66]

Animal (Age)	Exposure duration	Epigenetic mechanisms	Pathophysiological outcomes (possible)	References
Mice (PND20, 180 and 700)	PND1 to PND20	↑miR-106b at PND20 (1.5 fold) ↓ miR-34c at PND180 (1.6 fold) ↑miR-29b at PND20 (1.6 fold) ↑ miR-132 at PND20 (4.8 fold) ↓miR-124 at PND700 (2 fold)	Alzheimer's disease, tauopathies	[67]
Rats (20–22 days)	8 weeks	↑miR-211 with 300ppm Pb (1.75 old) ↓ miR-494 with 300ppm Pb (2.04 fold) ↑miR-449a with 300ppm Pb (2.89 fold) ↑miR-34c with 300ppm Pb (4.05 fold) ↑miR-34b with 300ppm Pb (4.48 fold) ↑miR-204 with 300ppm Pb (5.48 fold) ↑miR-448 with 300ppm Pb (30.51 fold) ↓mRNA with 300ppm Pb (Bcl2, Itpr1) ↓mRNA Map2k1 with 300ppm Pb	Neural injury, neurodegeneration, axon and synapse dysfunction, impaired neural development and regeneration, impaired performance, Alzheimer's disease, Parkinson's disease and depression	[68]
Rats (2-4 weeks and 12–14 weeks)	40D	Apoptotic cells with irregular nuclear membrane, chromatin clumping, and nuclear fragmentation	Apoptosis	[69]
Mice (10 months)	2 weeks prior to mating and continued throughout gestation to 3 weeks after birth	ARTN & C5aR1 methylation 32 ppm Pb exposure) Ankdd1b methylation 2.1 ppm Pb exposure)	Death of neurons, Alzheimer's disease, migraine, and major depressive disorder	[70]
Rat (PND55)	PERI: 10 days prior to breeding to PND 21 EPN: birth through weaning (PND 21) LPN: birth through postnatal day 55	Quantities of methylation changes at gene promotor region and varies according to genders	Schizophrenia, Alzheimer's disease, memory impairment, etc.	[5]
Mice (E18, PND0, PND6, and PND60)	2 months prior to breeding and throughout lactation	Changes of H3K9Ac, H3K4Me3, H3K9Me2, H3K27Me3 level	Cognition deficits, behavioral dysfunction, neurodevelopmental disorders	[71]

Animal (Age)	Exposure duration	Epigenetic mechanisms	Pathophysiological outcomes (possible)	References
Mice (PND0 and PND6)	Pb acetate for 2 months prior to breeding until sacrifice	Changes of H3K9/14Ac and H3K9Me3 level	Cognitive/behavioral problems during childhood	[63]
Mice (PND20 and PND50)	PND 1 to PND 20	↓MECP2 ↑Dnmt3a mRNA & miR-29b (PND50) ↓DNMT1 mRNA (PND50) ↑miR-148a(PND50) ↑SP1 mRNA(PND20) ↑miR-124(PND50) ↓APP mRNA(PND20) ↑miR-106b(PND50)	Tau-induced cell apoptosis in AD; neurodegeneration	[72]

Table 1.
Summary of some literature concerning epigenetic changes involved in lead neurotoxicity.

altered expression of objective genes in a gender-dependent manner. 5hmC may also serve as potential biomarkers for lead susceptibility to neurological diseases [76].

As a general rule, the changes of DNA methylation levels were often negatively correlated with the transcription levels of the objective genes. But this association was dependent on gender, the exposure periods, and their relative genetic locations. Interestingly, for females, genes regulated by DNA methylation were inclined to encode RNA- and protein-related processes; and for males, the enriched pathways included signaling pathways, stress, and neural responses to stimuli [5].

4.2 Histone modification

Compared to DNA methylation, fewer associations between histone modification and lead neurotoxicity were reported. Categorized by posttranslational forms, most studies focused on histone acetylation changes in response to lead exposure. A specific histone acetylation level results from the balanced counteraction of histone acetyltransferases (HATs) and histone deacetylases (HDACs). An acetylated form normally corresponds to a more relaxed chromatin status, leading to an enhanced expression of the target genes, and *vice versa* [77]. In 2014, our lab published a relatively novel article describing an increased acetylated form of histone H3 as exposed by 5 or 25 mg/l of lead. This alteration accompanies with the enhanced transcription of p300, a typical HAT [64]. Subsequently, the interesting point is that chronic lead exposure reduced the total level of H3K9ac, displaying an opposite tendency to total H3ac levels, which unveiled a specific alteration depending on the concrete acetylation sites, as well as the neural models and exposure conditions used [78].

Some innovative progresses were made in the primates with early life exposure. Bihaqi et al. observed that apart from DNMTs and MeCP2, lead also caused lifelong alterations of H3K9ac, H3K8ac, and H4K12ac, which levels were increased only in 23-year-old adults, not in 12-year-old primates [14]. In another instance, perinatal lead exposure downregulated H3K9ac levels in aging mice, a proof that key epigenetic regulators can be linked with development of Alzheimer's disease [15]. Murine hippocampus and frontal cortex were similar in lead-induced epigenetic changes, that is, H3K9/14ac was gradually reduced as exposure prolonged, factored by mixed actions of gender and prenatal stress [63]. The literature pertaining to histone modifications and lead neurotoxicity was shown in **Table 1**.

Different from histone acetylation, histone methylation gained a stricter site specificity and more stable to maintain gene expression patterns [79]. With the developmental exposure of lead, primates of 400-day olds were subjected to epigenetic examination in brains. H3K4me2 was found to be increased significantly, indicating an activated propensity of related gene expression [14]. In another animal study, H3K9me3 displayed a relatively stable tendency with treatment of lead in mice, and those cases varied depending on the studied brain regions and genders [63]. According to general knowledge gained in this field, H3K4me basically played roles in promoting gene expression, while H3K27me and H3K9me mostly displayed negative regulatory activity. Our previous finding underpinned the importance of H3K27me3 in modulating lead-induced spatial memory deficits [80]. We found that chronic lead exposure perinatally could reduce the global H3K27me3 levels in rat hippocampus, and this alteration led to a genome-wide reprogramming of this repressive epigenetic mark on the target genes. This result gave a picture of how an epigenetic change can give rise to a global genetic response and the ensuing adverse neurological outcomes.

In contrast with acetylation and methylation, very few studies were shown to investigate the interaction of lead neurotoxicity and other forms of histone modifications. In spite of deficiency of relevant literature, this study is supposed to be promising to totally decipher histone codes involved in lead neurotoxicity. H2A ubiquitination was recently found to be associated with DNA damage response, which suggested that site-specific histone ubiquitination organizes the spatiotemporal recruitment of DNA repair factors, and these recruitments facilitated DNA repair pathway choice between homologous recombination and nonhomologous end joining [81].

4.3 ncRNA

ncRNAs are epigenetic regulators susceptible to environmental signals. Among diverse forms of ncRNAs, microRNA (miRNA) was most extensively studied concerning their relations with lead neurotoxicity. In rats chronically exposed to lead, at least seven miRNAs were altered considering their expression levels. In details, miR-204, miR-211, miR-448, miR-449a, miR-34b, and miR34c were dramatically upregulated, while miR-494 was downregulated. These miRNAs were implicated in regulating genes involved in neurodegeneration, synaptogenesis, and neuronal injury [68]. Masoud et al. observed that early life lead exposure yielded a transient increase in the expression of AD-related miRNAs, such as miR-106b, miR-29b, and miR-132 [67]. Another rat exposure model with 100 ppm Pb also gave some evidence that some miRNAs, mainly targeting to a histone methyltransferase EZH2, were divergently regulated by lead in pup hippocampus. In response to lead, abundance of miR-137 and miR-101 was elevated, and miR-144d was decreased. The aberrant stimulation miR-137 may have important physiological relevance, as it formed a negative regulatory loop with EZH2, which drove the downregulation of H3K27me3 [80]. Some examples of miRNA changes during lead-induced neurotoxicity were shown in **Table 1**.

It is insufficient for other forms of ncRNAs remodeled by lead exposure in CNS. In 2018, Nan published an interesting article with relevance to this research field [82]. The authors identified a novel lincRNA (long noncoding RNA), namely lncRNAL20992, as a key responsor toward lead neurotoxicity. lncRNAL20992 was significantly upregulated in a lead-induced neuronal injury model. Four proteins were found to physically interact with lncRNAL20992 to mediate the lead-induced neuronal injury. To date, few associations of piRNAs or circRNAs were discovered with lead-induced neurotoxicity. Interestingly, a 98-nucleotide nuclear RNA with unknown function called Rn4.5s, as well as a RNA precursor Rn45s, showed changes in methylation in the hippocampus of females exposed to 3 ppm of lead [11].

5. Epigenetic mechanisms of lead-induced Alzheimer's disease

By modifying the global epigenetic landscape, early life lead exposure had not only immediate adverse consequences for brain development but also persistent effects till the later life. This toxic property may increase the susceptibility of organisms to diseases, especially CNS-related diseases [3]. It has been long established that chronic lead contact is an important risk factor for the pathogenesis of Alzheimer's disease [83]. This finding is significant because AD is the most common form of dementia, which affects aging individual. There were at least 25 million people worldwide affected by this disease in 2003, including at least 4.5 million people in the United States [84]. Similar to other neurodegenerative diseases, AD is a complex and heterogeneous disorder with both environmental and genetic etiology.

There are a variety of molecular mechanisms proposed to mediate the lead-induced pathogenesis of AD. In monkeys, Wu et al. discovered that with lead exposure in infants, the aging monkeys exhibited abnormal expression alterations of AD-related genes and a key transcriptional regulator specificity protein 1 (Sp1). This was manifested by increasing AβPP, b-site AβPP cleaving enzyme (BACE), and Aβ and by decreasing DNA methyltransferase activity [16]. In an epidemiological review toward the Mexican population, it was summarized that early-life lead exposure was a potential risk factor for AD in the Mexican population [85]. While these mechanisms partly explained the key cellular and molecular changes brought by lead exposure, it is still pivotal to figure out the "fetal programming" phenomenon involved in the studied pathogenesis.

Roles of epigenetics are much appreciated due to their similar modes of action with "fetal origin of adult disease," which characterizes the basic regularities underlying lead-induced pathogenesis of AD [3, 11]. One anticipated way to achieve long-lasting or permanent changes in gene expression is to alter the structural makeup of the DNA bases that led to hypermethylation or hypomethylation consequences. The changes of DNA methylation in promoters or other gene regulatory components are found to be extremely stable and can even transmitted to the next generation. This style of action is consistent with the neurotoxic course of Pb to induce AD and is anticipated to fetal program the key AD-related genes, enabling their long abnormal transcriptions. Compared to CpG methylation, histone modifications are normally unable to elicit permanent regulatory effect till aging period, as evidenced by previous findings that H3K9ac and H3K4me2 followed lifespan-dependent changing curves, whereas variable time points showed different altering orientations [15]. Therefore, CpG methylation might be a competent epigenetic mechanism of choice to be used to explain the long development of AD. In another aspect, susceptible response from specific histone modification may be used to early predict the risk of AD, on the basis of their potential genetic associations.

Given these observations, White et al. proposed the possible pathways by which epigenetic factors mediate Pb-led pathogenesis of AD (**Figure 2**) [25]. They regarded epigenetics, mainly represented by CpG methylation, as potential key mechanisms that manifested delay consequences of lead exposure and consequent Alzheimer's disease. According to this hypothesis, early-life lead exposure caused a cascade of molecular changes exemplified by reducing CpG methylations at promoters of key AD-related genes, such as APP and BACE. Assisted by the actions of MeCP2 or SP1, these methylation reductions promoted the gene expression toward the corresponding proteins. Subsequently, Aβ was synthesized and started to accumulate in the later life. When the accumulation reached to a threshold value,

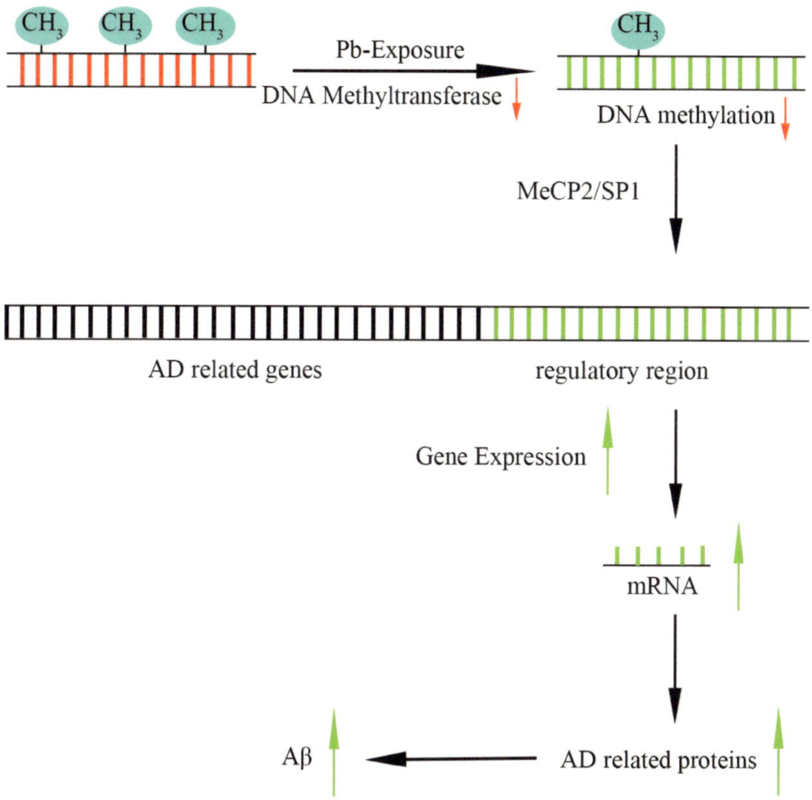

Figure 2.
Possible epigenetic mechanism of Alzheimer's disease induced by early life lead exposure.

AD symptoms started to emerge and progress. This hypothesis gives an intriguing example to implicate epigenetic factors with Alzheimer's disease induced by early-life lead exposure.

6. Conclusion and future directions

In conclusion, epigenetic factors played essential roles in mediating lead-induced neurotoxicity. Comprehensive investigations unveiled the importance of CpG methylations in multiple genetic loci in rodents and primates. CpG methylation on a specific gene promoter might give rise to a long-term suppression of gene expression, in which case formed a phenomenon of "fetal programming" of neurological disease. In addition, changes of histone modifications might reflect a relatively dynamic signal to moderate the ensuing molecular relay and neurotoxic manifestations. As a newly emerging research field, it is anticipated to have several future directions about relevant of epigenetic factors to lead neurotoxicity: (1) new epigenetic mechanisms, such as 5hmC, RNA methylation, and scarcely mentioned ncRNA forms, need to be thoroughly investigated regarding their associations with lead neurotoxicity; (2) most previous studies observed the huge impact of gender on the neurotoxic performances, but very few explanations were provided. Epigenetic differences and causing agents between the genders should be investigated with insight; (3) there are currently some medicine developed to target epigenetic sites, like HDAC inhibitors. However, the sole histone deacetylase is unable to reflect the global epigenetic aspects and to use for intervention with full efficacy. Other specific reversing pharmaceuticals targeting epigenetic factors are warranted to be developed to interfere with development of neurological disease induced by lead exposure.

Acknowledgements

The work of this chapter was supported by the Fundamental Research Funds for the Central Universities (JZ2020HGTB0053) and the Key Research and Development Project in Anhui (201904e01020001).

Conflict of interest

The authors declare no conflict of interest.

Abbreviations

5hmC	5-hydroxymethylated cytosine
AD	Alzheimer's disease
BACE	b-site AβPP cleaving enzyme
BDNF	brain-derived neurotrophic factor
BLL	blood lead level
CDC	Centers for Disease Control and Prevention
CNS	central nervous system
CpG	Cytosine Guanine dinucleotides
DAT1	dopamine transporter 1
DMR	Differentially Methylated Region
DNMT	DNA methyltransferases
DRD4	dopamine receptor 4
EPSC	excitatory postsynaptic currents
GABA	γ-aminobutyric acid
H3K8ac	lysine acetylation at histone H3K8
H3K9ac	lysine acetylation at histone H3K9
H4K12ac	lysine acetylation at histone H4K12
H3K4me	lysine methylation at histone H3K4
H3K9me	lysine methylation at histone H3K9
H3K27me	lysine methylation at histone H3K27
HAT	histone acetyltransferases
HDAC	histone deacetylase
lncRNA	long noncoding RNA
IPSC	inhibitory postsynaptic currents
MeCP2	methyl-cytosine-phosphate-guanine (Me-CpG) binding protein-2
NMDAR	N-methyl-D-aspartate receptor
ncRNA	noncoding RNA
Pb	lead
piRNA	piwi-interacting RNA
PTM	posttranslational modifications
snoRNA	small nucleolar RNA
SP1	regulator specificity protein 1

Author details

Yi Xu*, Tian Wang and Jie Zhang
School of Food Science and Bioengineering, Hefei University of Technology, Hefei, China

*Address all correspondence to: xuyixuyi3734@163.com

References

[1] Neal AP, Guilarte TR. Molecular neurobiology of lead (Pb^{2+}): Effects on synaptic function. Molecular Neurobiology. 2010;**42**(3):151-160

[2] Raciti M, Ceccatelli S. Epigenetic mechanisms in developmental neurotoxicity. Neurotoxicology and Teratology. 2018;**66**:94-101

[3] Senut MC, Cingolani P, Sen A, Kruger A, Shaik A, Hirsch H, et al. Epigenetics of early-life lead exposure and effects on brain development. Epigenomics. 2012;**4**(6):665-674

[4] Senut MC, Sen A, Cingolani P, Shaik A, Land SJ, Ruden DM. Lead exposure disrupts global DNA methylation in human embryonic stem cells and alters their neuronal differentiation. Toxicological Sciences. 2014;**139**(1):142-161

[5] Singh G, Singh V, Wang ZX, Voisin G, Lefebvre F, Navenot JM, et al. Effects of developmental lead exposure on the hippocampal methylome: Influences of sex and timing and level of exposure. Toxicology Letters. 2018;**290**:63-72

[6] Eid A, Zawia N. Consequences of lead exposure, and it's emerging role as an epigenetic modifier in the aging brain. Neurotoxicology. 2016;**56**:254-261

[7] https://www.cdc.gov/nceh/lead/data/index.htm [Internet]. 2020

[8] Neal AP. Mechanisms of heavy metal neurotoxicity: Lead and manganese. Journal of Drug Metabolism & Toxicology. 2015;**06**(03). DOI: 10.4172/2157-7609.S5-002

[9] Bellinger DC. Very low lead exposures and children's neurodevelopment. Current Opinion in Pediatrics. 2008; **20**(2):172-177

[10] Sanders T, Liu Y, Buchner V, Tchounwou PB. Neurotoxic effects and biomarkers of lead exposure: A review. Reviews on Environmental Health. 2009;**24**(1):15-45

[11] Sanchez-Martin FJ, Lindquist DM, Landero-Figueroa J, Zhang X, Chen J, Cecil KM, et al. Sex- and tissue-specific methylome changes in brains of mice perinatally exposed to lead. Neurotoxicology. 2015;**46**:92-100

[12] Weuve J, Press DZ, Grodstein F, Wright RO, Hu H, Weisskopf MG. Cumulative exposure to lead and cognition in persons with Parkinson's disease. Movement Disorders. 2013;**28**(2):176-182

[13] Wu S, Liu H, Zhao H, Wang X, Chen J, Xia D, et al. Environmental lead exposure aggravates the progression of Alzheimer's disease in mice by targeting on blood brain barrier. Toxicology Letters. 2020;**319**:138-147

[14] Bihaqi SW, Huang H, Wu J, Zawia NH. Infant exposure to lead (Pb) and epigenetic modifications in the aging primate brain: Implications for Alzheimer's disease. Journal of Alzheimer's Disease. 2011;**27**(4):819-833

[15] Eid A, Bihaqi SW, Renehan WE, Zawia NH. Developmental lead exposure and lifespan alterations in epigenetic regulators and their correspondence to biomarkers of Alzheimer's disease. Alzheimer's & Dementia. 2016;**2**:123-131

[16] Wu J, Basha MR, Brock B, Cox DP, Cardozo-Pelaez F, McPherson CA, et al. Alzheimer's disease (AD)-like pathology in aged monkeys after infantile exposure to environmental metal lead (Pb): Evidence for a developmental origin and environmental link for AD. The Journal of Neuroscience. 2008;**28**(1):3-9

[17] Mazumdar M, Xia W, Hofmann O, Gregas M, Ho Sui S, Hide W, et al. Prenatal lead levels, plasma amyloid beta levels, and gene expression in young adulthood. Environmental Health Perspectives. 2012;**120**(5):702-707

[18] Park SS, Skaar DA, Jirtle RL, Hoyo C. Epigenetics, obesity and early-life cadmium or lead exposure. Epigenomics. 2017;**9**(1):57-75

[19] Goodrich JM, Sanchez BN, Dolinoy DC, Zhang Z, Hernandez-Avila M, Hu H, et al. Quality control and statistical modeling for environmental epigenetics: A study on in utero lead exposure and DNA methylation at birth. Epigenetics. 2015;**10**(1):19-30

[20] Lidsky TI, Schneider JS. Lead neurotoxicity in children: Basic mechanisms and clinical correlates. Brain. 2003;**126**(Pt 1):5-19

[21] Wilson MA, Johnston MV, Goldstein GW, Blue ME. Neonatal lead exposure impairs development of rodent barrel field cortex. Proceedings of the National Academy of Sciences of the United States of America. 2000;**97**(10):5540-5545

[22] Chen C, Tonegawa S. Molecular genetic analysis of synaptic plasticity, activity-dependent neural development, learning, and memory in the mammalian brain. Annual Review of Neuroscience. 1997;**20**:157-184

[23] Nihei MK, Desmond NL, McGlothan JL, Kuhlmann AC, Guilarte TR. N-methyl-D-aspartate receptor subunit changes are associated with lead-induced deficits of long-term potentiation and spatial learning. Neuroscience. 2000;**99**(2):233-242

[24] Zhang XY, Liu AP, Ruan DY, Liu J. Effect of developmental lead exposure on the expression of specific NMDA receptor subunit mRNAs in the hippocampus of neonatal rats by digoxigenin-labeled in situ hybridization histochemistry. Neurotoxicology and Teratology. 2002;**24**(2):149-160

[25] White LD, Cory-Slechta DA, Gilbert ME, Tiffany-Castiglioni E, Zawia NH, Virgolini M, et al. New and evolving concepts in the neurotoxicology of lead. Toxicology and Applied Pharmacology. 2007;**225**(1):1-27

[26] Lasley SM, Gilbert ME. Rat hippocampal glutamate and GABA release exhibit biphasic effects as a function of chronic lead exposure level. Toxicological Sciences. 2002;**66**(1):139-147

[27] Braga MF, Pereira EF, Albuquerque EX. Nanomolar concentrations of lead inhibit glutamatergic and GABAergic transmission in hippocampal neurons. Brain Research. 1999;**826**(1):22-34

[28] Ding JJ, Zou RX, He HM, Lou ZY, Xu Y, Wang HL. Pb inhibits hippocampal synaptic transmission via cyclin-dependent kinase-5 dependent Synapsin 1 phosphorylation. Toxicology Letters. 2018;**296**:125-131

[29] Matsuda N, Lu H, Fukata Y, Noritake J, Gao H, Mukherjee S, et al. Differential activity-dependent secretion of brain-derived neurotrophic factor from axon and dendrite. The Journal of Neuroscience. 2009;**29**(45):14185-14198

[30] Neal AP, Stansfield KH, Worley PF, Thompson RE, Guilarte TR. Lead exposure during synaptogenesis alters vesicular proteins and impairs vesicular release: Potential role of NMDA receptor-dependent BDNF signaling. Toxicological Sciences. 2010;**116**(1):249-263

[31] Walz C, Jungling K, Lessmann V, Gottmann K. Presynaptic plasticity

in an immature neocortical network requires NMDA receptor activation and BDNF release. Journal of Neurophysiology. 2006;**96**(6):3512-3516

[32] Du Y, Ge MM, Xue W, Yang QQ, Wang S, Xu Y, et al. Chronic Lead exposure and mixed factors of GenderxAgexBrain regions interactions on dendrite growth, spine maturity and NDR kinase. PLoS One. 2015;**10**(9):e0138112

[33] Bihaqi SW. Early life exposure to lead (Pb) and changes in DNA methylation: Relevance to Alzheimer's disease. Reviews on Environmental Health. 2019;**34**(2):187-195

[34] Sallmen M, Suvisaari J, Lindbohm ML, Malaspina D, Opler MG. Paternal occupational lead exposure and offspring risks for schizophrenia. Schizophrenia Research. 2016;**176**(2-3): 560-565

[35] Perera F, Herbstman J. Prenatal environmental exposures, epigenetics, and disease. Reproductive Toxicology. 2011;**31**(3):363-373

[36] Bannister AJ, Kouzarides T. Regulation of chromatin by histone modifications. Cell Research. 2011;**21**(3):381-395

[37] Esteller M. Non-coding RNAs in human disease. Nature Reviews. Genetics. 2011;**12**(12):861-874

[38] Nye MD, Fry RC, Hoyo C, Murphy SK. Investigating epigenetic effects of prenatal exposure to toxic metals in newborns: Challenges and benefits. Medical Epigenetics. 2014;**2**(1):53-59

[39] Xu Y, Chen XT, Luo M, Tang Y, Zhang G, Wu D, et al. Multiple epigenetic factors predict the attention deficit/hyperactivity disorder among the Chinese Han children. Journal of Psychiatric Research. 2015;**64**:40-50

[40] Campos EI, Reinberg D. Histones: Annotating chromatin. Annual Review of Genetics. 2009;**43**:559-599

[41] Farrelly LA, Maze I. An emerging perspective on 'histone code' mediated regulation of neural plasticity and disease. Current Opinion in Neurobiology. 2019;**59**:157-163

[42] Bhanu NV, Sidoli S, Yuan ZF, Molden RC, Garcia BA. Regulation of proline-directed kinases and the trans-histone code H3K9me3/H4K20me3 during human myogenesis. The Journal of Biological Chemistry. 2019;**294**(20):8296-8308

[43] Kao AC-C, Chan KW, Anthony DC, Lennox BR, Burnet PW. Prebiotic reduction of brain histone deacetylase (HDAC) activity and olanzapine-mediated weight gain in rats, are acetate independent. Neuropharmacology. 2019;**150**:184-191

[44] Bagheri A, Habibzadeh P, Razavipour SF, Volmar C-H, Chee NT, Brothers SP, et al. HDAC inhibitors induce expression and promote neurite outgrowth in human neural progenitor cells-derived neurons. International Journal of Molecular Sciences. 2019;**20**(5):1109-1124

[45] Magri F, Vanoli F, Corti S. miRNA in spinal muscular atrophy pathogenesis and therapy. Journal of Cellular and Molecular Medicine. 2018;**22**(2):755-767

[46] Smrt RD, Szulwach KE, Pfeiffer RL, Li X, Guo W, Pathania M, et al. MicroRNA miR-137 regulates neuronal maturation by targeting ubiquitin ligase mind bomb-1. Stem Cells. 2010;**28**(6):1060-1070

[47] Fuks F. DNA methylation and histone modifications: Teaming up to silence genes. Current Opinion in Genetics & Development. 2005;**15**(5):490-495

[48] Honda S, Bicocca VT, Gessaman JD, Rountree MR, Yokoyama A, Yu EY,

et al. Dual chromatin recognition by the histone deacetylase complex HCHC is required for proper DNA methylation in *Neurospora crassa*. Proceedings of the National Academy of Sciences of the United States of America. 2016;**113**(41):E6135-E6E44

[49] Shahbazian MD, Zoghbi HY. Rett syndrome and MeCP2: Linking epigenetics and neuronal function. American Journal of Human Genetics. 2002;**71**(6):1259-1272

[50] Azpurua J, Eaton BA. Neuronal epigenetics and the aging synapse. Frontiers in Cellular Neuroscience. 2015;**9**:208

[51] Li Y, Xie C, Murphy SK, Skaar D, Nye M, Vidal AC, et al. Lead exposure during early human development and DNA methylation of imprinted gene regulatory elements in adulthood. Environmental Health Perspectives. 2016;**124**(5):666-673

[52] Vandegehuchte MB, Janssen CR. Epigenetics and its implications for ecotoxicology. Ecotoxicology. 2011;**20**(3):607-624

[53] Mani MS, Kabekkodu SP, Joshi MB, Dsouza HS. Ecogenetics of lead toxicity and its influence on risk assessment. Human & Experimental Toxicology. 2019;**38**(9):1031-1059

[54] Sanchez OF, Lee J, Yu King Hing N, Kim SE, Freeman JL, Yuan C. Lead (Pb) exposure reduces global DNA methylation level by non-competitive inhibition and alteration of DNMT expression. Metallomics. 2017;**9**(2):149-160

[55] Pilsner JR, Hu H, Ettinger A, Sanchez BN, Wright RO, Cantonwine D, et al. Influence of prenatal lead exposure on genomic methylation of cord blood DNA. Environmental Health Perspectives. 2009;**117**(9):1466-1471

[56] Dosunmu R, Alashwal H, Zawia NH. Genome-wide expression and methylation profiling in the aged rodent brain due to early-life Pb exposure and its relevance to aging. Mechanisms of Ageing and Development. 2012;**133**(6):435-443

[57] Schneider JS, Kidd SK, Anderson DW. Influence of developmental lead exposure on expression of DNA methyltransferases and methyl cytosine-binding proteins in hippocampus. Toxicology Letters. 2013;**217**(1):75-81

[58] Levenson JM, Roth TL, Lubin FD, Miller CA, Huang IC, Desai P, et al. Evidence that DNA (cytosine-5) methyltransferase regulates synaptic plasticity in the hippocampus. The Journal of Biological Chemistry. 2006;**281**(23):15763-15773

[59] Wu S, Hivert MF, Cardenas A, Zhong J, Rifas-Shiman SL, Agha G, et al. Exposure to low levels of Lead in utero and umbilical cord blood DNA methylation in project viva: An epigenome-wide association study. Environmental Health Perspectives. 2017;**125**(8):087019

[60] Montrose L, Faulk C, Francis J, Dolinoy DC. Perinatal lead (Pb) exposure results in sex and tissue-dependent adult DNA methylation alterations in murine IAP transposons. Environmental and Molecular Mutagenesis. 2017;**58**(8):540-550

[61] Alashwal H, Dosunmu R, Zawia NH. Integration of genome-wide expression and methylation data: Relevance to aging and Alzheimer's disease. Neurotoxicology. 2012;**33**(6):1450-1453

[62] Duan Y, Peng L, Shi H, Jiang Y. The effects of lead on GABAergic interneurons in rodents. Toxicology and Industrial Health. 2017;**33**(11):867-875

[63] Schneider JS, Anderson DW, Kidd SK, Sobolewski M, Cory-Slechta DA. Sex-dependent effects of lead and prenatal stress on post-translational histone modifications in frontal cortex and hippocampus in the early postnatal brain. Neurotoxicology. 2016;**54**:65-71

[64] Luo M, Xu Y, Cai R, Tang Y, Ge MM, Liu ZH, et al. Epigenetic histone modification regulates developmental lead exposure induced hyperactivity in rats. Toxicology Letters. 2014;**225**(1):78-85

[65] Hill DS, Cabrera R, Wallis Schultz D, Zhu H, Lu W, Finnell RH, et al. Autism-like behavior and epigenetic changes associated with autism as consequences of in utero exposure to environmental pollutants in a mouse model. Behavioural Neurology. 2015;**2015**:426263

[66] Dash M, Eid A, Subaiea G, Chang J, Deeb R, Masoud A, et al. Developmental exposure to lead (Pb) alters the expression of the human tau gene and its products in a transgenic animal model. Neurotoxicology. 2016;**55**:154-159

[67] Masoud AM, Bihaqi SW, Machan JT, Zawia NH, Renehan WE. Early-life exposure to lead (Pb) alters the expression of microRNA that target proteins associated with Alzheimer's disease. Journal of Alzheimer's Disease. 2016;**51**(4):1257-1264

[68] An J, Cai T, Che H, Yu T, Cao Z, Liu X, et al. The changes of miRNA expression in rat hippocampus following chronic lead exposure. Toxicology Letters. 2014;**229**(1):158-166

[69] Sharifi AM, Mousavi SH, Jorjani M. Effect of chronic lead exposure on pro-apoptotic Bax and anti-apoptotic Bcl-2 protein expression in rat hippocampus in vivo. Cellular and Molecular Neurobiology. 2010;**30**(5):769-774

[70] Dou JF, Farooqui Z, Faulk CD, Barks AK, Jones T, Dolinoy DC, et al. Perinatal lead (Pb) exposure and cortical neuron-specific DNA methylation in male mice. Genes (Basel). 2019;**10**(4):274-288

[71] Varma G, Sobolewski M, Cory-Slechta DA, Schneider JS. Sex- and brain region-specific effects of prenatal stress and lead exposure on permissive and repressive post-translational histone modifications from embryonic development through adulthood. Neurotoxicology. 2017;**62**:207-217

[72] Masoud AM, Bihaqi SW, Alansi B, Dash M, Subaiea GM, Renehan WE, et al. Altered microRNA, mRNA, and protein expression of neurodegeneration-related biomarkers and their transcriptional and epigenetic modifiers in a human tau transgenic mouse model in response to developmental lead exposure. Journal of Alzheimer's Disease. 2018;**63**(1):273-282

[73] Khalid M, Abdollahi M. Epigenetic modifications associated with pathophysiological effects of lead exposure. Journal of Environmental Science and Health. Part C, Environmental Carcinogenesis & Ecotoxicology Reviews. 2019;**37**(4): 235-287

[74] Kundakovic M, Lim S, Gudsnuk K, Champagne FA. Sex-specific and strain-dependent effects of early life adversity on behavioral and epigenetic outcomes. Frontiers in Psychiatry. 2013;**4**:78

[75] Szulwach KE, Li X, Li Y, Song CX, Wu H, Dai Q, et al. 5-hmC-mediated epigenetic dynamics during postnatal neurodevelopment and aging. Nature Neuroscience. 2011;**14**(12):1607-1616

[76] Sen A, Cingolani P, Senut MC, Land S, Mercado-Garcia A, Tellez-Rojo MM, et al. Lead exposure induces changes in 5-hydroxymethylcytosine

clusters in CpG islands in human embryonic stem cells and umbilical cord blood. Epigenetics. 2015;**10**(7):607-621

[77] Hutson TH, Kathe C, Palmisano I, Bartholdi K, Hervera A, De Virgiliis F, et al. Cbp-dependent histone acetylation mediates axon regeneration induced by environmental enrichment in rodent spinal cord injury models. Science Translational Medicine. 2019;**11**(487):2064-2076

[78] Wu Y, Xu Y, Huang X, Ye D, Han M, Wang HL. Regulatory roles of histone deacetylases 1 and 2 in Pb-induced neurotoxicity. Toxicological Sciences. 2018;**162**(2):688-701

[79] Shafabakhsh R, Aghadavod E, Ghayour-Mobarhan M, Ferns G, Asemi Z. Role of histone modification and DNA methylation in signaling pathways involved in diabetic retinopathy. Journal of Cellular Physiology. 2019;**234**(6):7839-7846

[80] Gu X, Xu Y, Xue WZ, Wu Y, Ye Z, Xiao G, et al. Interplay of miR-137 and EZH2 contributes to the genome-wide redistribution of H3K27me3 underlying the Pb-induced memory impairment. Cell Death & Disease. 2019;**10**(9):671

[81] Uckelmann M, Sixma TK. Histone ubiquitination in the DNA damage response. DNA Repair. 2017;**56**:92-101

[82] Nan A, Jia Y, Li X, Liu M, Zhang N, Chen L, et al. Editor's highlight: lncRNAL20992 regulates apoptotic proteins to promote lead-induced neuronal apoptosis. Toxicological Sciences. 2018;**161**(1):115-124

[83] Basha MR, Wei W, Bakheet SA, Benitez N, Siddiqi HK, Ge YW, et al. The fetal basis of amyloidogenesis: Exposure to lead and latent overexpression of amyloid precursor protein and beta-amyloid in the aging brain. The Journal of Neuroscience. 2005;**25**(4):823-829

[84] Hebert LE, Scherr PA, Bienias JL, Bennett DA, Evans DA. Alzheimer disease in the US population: Prevalence estimates using the 2000 census. Archives of Neurology. 2003;**60**(8):1119-1122

[85] Chin-Chan M, Cobos-Puc L, Alvarado-Cruz I, Bayar M, Ermolaeva M. Early-life Pb exposure as a potential risk factor for Alzheimer's disease: Are there hazards for the Mexican population? Journal of Biological Inorganic Chemistry. 2019;**24**(8):1285-1303

Chapter 3

Lead: Toxicological Profile, Pollution Aspects and Remedial Solutions

Rajendra Sukhadeorao Dongre

Abstract

Water quality is the keen concern all over the world. As water resources get contaminated naturally or artificially, at last they affect our health as well as economic and social development of the nations. Among the prominent chemical pollutants, lead alone has threatened health of billions due to deterioration in water quality. Thus, safe water for drinking becomes a major worried issue of UNICEF as a major stakeholder in sustaining water quality with responsibility to improve and sustain quality of water through its programs around the world. The mitigation of contaminations through effective techniques is the common global effort toward remediation of pollutions. This chapter has investigated viability of ordinary organic waste and natural polymeric material like chitin, chitosan, and doped bio-composite adsorbents for the mitigation of Pb^{2+} ions from water. The relatively higher adsorption potential of synthesized composites is studied for removing excessive lead concentration from water as tested in batch and column mode. Certain exchangeable metal ions such as Na^{+}, Ca^{2+}, and Mg^{2+} aid to form solid precipitate, which is effective in Pb^{2+} retention and facilitate its adsorption. Certain pilot domestic units provide simple, efficient, and fesasible options for removal of Pb^{2+} ions from water.

Keywords: lead(II), chitosan, biocomposite, adsorption, diffusion, thermodynamic parameter, batch/column modes

1. Introduction

The need of pure and safe water to the society is a precondition for health and development along with a basic human right as very well cited in the slogan below:

Water is abundant, besides inexpensive, Yet pure Water is fewer,
While good health, needs to drink it in safe, pure and upmost care!

Yet, millions of people throughout developing world denied this fundamental fact/right. Water borne diseases may arise through insufficient safe water supplies coupled with poor sanitation, and hygiene causes about 4 million death/year including children [1]. International communities are doing lots of attempts in this sense. Billion people do not have access to improve water sources and are not able to get pure and safe water for drinking as well as for other purposes. In fact, the magnitude of water quality problem is even larger than being projected. The existing improved sources in developing countries do not provide water of adequate quality for drinking and other purposes.

In 2006, UNICEF has drawn global strategy to distinguish growing importance of ensuring safe drinking water, sanitation, and hygiene strategies for the decade 2006–2015 in its special programs [1–3]. In many Asian countries, the tube well water is viable for extensive toxic and hazardous contaminations such as arsenic, fluoride, and lead [1, 2]. Even today more than half of the world population depends on ground water being a vital source for drinking as it contributes 97% of global freshwater. In a number of geographical regions, ground water is a vital source for drinking due to the fact that it contributes single largest supply for serving drinking water. Especially in India, almost 80% ground water is needed for rural domestic need and 50% needed for urban population. Thus, clean ground water supply is most essential to serve the basic and critical necessities for various utility purposes. Over the years, the availability of pure and safe waters for drinking and other functions has been great concern for better environment [3, 4].

Efficient synthetic purification/treatment processes are used to remove detrimental chemical/biological pollutants along with suspended solids from water and to provide pure and safe water for multiple consumptions such as drinking, therapeutics, pharmaceuticals, chemical, and industrial uses. Throughout the world many, water treatments are almost remains same as mention below:

i. physical processes like filtration, sedimentation, and distillation;

ii. biological methods such as slow sand filter/bioactive carbon; and

iii. chemical techniques, viz. coagulation flocculation, chlorination, and UV radiations.

Ground water is purer and safer than surface water due to earth covering that works as a natural filter, which is the major source for domestic purposes in the developing country like India [2, 5]. In fact, few decades back, water seems to be odorless, free from turbidity, and good from the esthetic point of view and considered to be pure and unpolluted. The concept of water pollution has now changed as even clear water may be latently polluted. As well, surface water is also contaminated by effluents from industries, municipalities, and other places. The contamination of hazardous anions like fluoride in particular and other anions such as nitrate, sulfate, and phosphate in ground water is a wide spread phenomenon causing health problem. The available domestic water purification processes are seldom suitable for rural people due to high cost and maintenance. This creates a great gap between developed water purification technology and its anticipated application [1–4, 6, 7].

Heavy metals such as Zn, Cu, As, Hg, Cr, Ni, Cd, and Pb cause severe problem for humans and aquatic ecosystems if discharged in water through industries and other sources due to particular toxic, hazardous, and carcinogenic nature as well as accumulations in the body based on relative chemical and physiological characteristics [1–4, 6, 8]. The removal of heavy metals from water is seriously needed because of the imposed environmental pollution and ecological degradation. The adsorption process is widely used for the removal of heavy metals from wastewater because of its low cost, availability, and eco-friendly nature. Both commercial adsorbents and bio-adsorbents are used for the removal of heavy metals from wastewater, with high removal capacity. This chapter compiles information on different adsorbents used for Pb^{2+} removal and provides information on the commercially available. Innovative processes and technologies are involved for lessening toxicity so as to cater demands of environmental standards being developed for the treatment of heavy metal polluted wastewater.

Toxic heavy metal removal from domestic wastes, sewage, and industrial outlets is a challenging task, especially in waste effluents. Various methodologies are being investigated as and when for the mitigation of such heavy metal pollution in water [1–4, 6, 8]. The heavy metals including lead, cadmium, and mercury are removed from industrial wastewater by means of assorted techniques namely physicochemical precipitation, electrochemical reduction, ion exchangers, reverse osmosis, cementation, electrodialysis, electrowinning, electrocoagulation, membrane separation, and adsorption. However, technical applications, plant easiness, and cost are vital parameters to choose utmost appropriate treatments for the mitigation of heavy metal pollution from water [4]. Physicochemical adsorption is quite a cheap and capable method for retention of heavy metals from industrial effluents due to its easy, successful, and profitable features. Heavy metal pollution subsists in wastes of a number of industries, such as metal plating, mining, tanneries, chloralkali, radiator built-up, smelting, alloy-making printed circuit board making, and storage batteries. Assorted metals such as tin, lead, and nickel metal-based solder plates shown familiar resistant over plated if applied for solderable applications as metal plating are recommended, as deposited inter-metallic formations ensuing diffusion barrier viable as good solder base for soldering applications. Most of the aforesaid industries discharged huge wastewater contaminated with lead and deemed utmost hazardous to man-health and environment conditions [1–4, 6].

Lead gets absorbed or accumulated in living species and subsequently penetrates in human systems through food chains/cycles. Ingested lead beyond its stringent/permitted level results in serious health disorders. Thus, it is compulsory to treat Pb^{2+} contaminations prior to its discharge into the environment [6, 8]. In fact, heavy metal lead is soft, malleable, bluish gray color being picky interested due to innate toxicity and extensive existence in the atmosphere. Lead is the most toxic metal considered as a priority pollutant as an industrial pollutant, which enters in an environment via soil, air, and water/wastewater. Lead is a systemic poison very toxic in nature because it causes anemia, kidney malfunction, brain tissue damage, and death in severe poisoning [9]. Lead pollution dispersed over the soil and ground water through natural sources and industrial effluents. Certain processing industries, such as acid-battery making, metal plating and finishing, ammunition, anti-knocking agent-tetraethyl-lead synthesis, ceramic/glass industries, and environmental clearout practices, dispersed lead polluted water, which is the foremost lead pollution sources. Alternatively, high lead concentration in the atmosphere also results in enduring health risks to all the elements of ecosystem.

As recommended by the WHO, the maximum tolerable limit of lead metal in drinking water is 0.05 mg/L, while Environmental Protection Agency allows the permissible limit for Pb^{2+} in wastewater as 50 ppb (part per billion). However, industrial wastewaters own Pb^{2+} ion with the amount ranging between 200 and 500 mg/L, which is a lot more than that of water quality standards. Maximum contamination/limiting level concentration and types of heavy metals as discharged are recognized by the USEPA as mentioned in **Table 1**.

1.1 Precedence pollutants in water

The introduction of contaminants into the environment causes pollution, which is the unfavorable alterations of our surroundings. This pollution can be in the form of chemical substances or energy such as noise, heat, and light [6]. Thus, the pollutants can be naturally occurring substances or energies; however, things are considered contaminants if they exist excess of natural levels. All such pollutants that enter the atmosphere by any known or unknown sources surely create harms or discomfort to humans and other living organisms of the planet.Many water

Entry	Heavy metal	Maximum contamination level (ppm)	Potential toxicity profiles
1	Lead	0.005	Damage/fatal to brain, kidney diseases, circulatory and nervous system disorders, death in severe contamination
2	Arsenic	0.050	Skin manifestations, visceral cancers, vascular disease
3	Cadmium	0.01	Kidney damage, renal disorder, human carcinogen
4	Chromium (VI)	0.05	Headache, diarrhea, nausea, vomiting, carcinogenic, respiratory tract problems
5	Copper	0.25	Wilson disease, insomnia, vomiting, hematemesis, hypotension, melena, coma, jaundice, pigmentation of skin, gastrointestinal distress, damage liver kidney
6	Nickel	0.20	Dermatitis, nausea, chronic asthma, coughing, human carcinogen
7	Mercury	0.00003	Rheumatoid arthritis, diseases of the kidneys, circulatory system, and nervous system

Table 1.
Maximum contamination/limiting level (MCL) concentration and types of heavy metals being discharged as recognized by the USEPA standards [1–4, 6].

pollutants may get discharged through domestic, industrial, and agricultural wastes due to human activities or domestic sources. In past years, water with odorless, colorless, and free from turbidity was considered as good, pure, and unpolluted from the esthetic point of view [1–4, 6]. But this entire concept of water pollution was changed. Now even if water is colorless and odorless, it can be polluted and may contain microbes as well as dissolved impurities such as toxic metals, organic pollutants, and radioactive materials. Thus, water pollution can be defined as any change in physical, chemical, or biological properties of matter. It is not possible to analysis water quality for all of chemical pollutants that could cause health problems, nor is it indispensable. However, most of these contaminants occur rarely, and many result from human contagion and affect a few water sources. The chemicals that potentially cause serious health problems by over widespread areas include arsenic, fluoride, phosphate, and nitrate (from swage and fertilizers) [1, 8].

Some of these anionic chemical pollutants are more often found in ground water, though surface water can also be impacted. In order to plan new water supply projects, especially to ground water resources, contamination from lead, arsenic, fluoride, phosphate, and nitrate should be given priority [5, 7, 8, 10]. Subsequently, second priority is to be given to metals (principally iron and manganese). The inorganic chemicals are common cause for water to be rejected from the esthetic view and also enhance salinity of water (saltiness or dissolved salt content of water body). The fresh or naturally occurring water has salinity $<0.05\%$ as characterized by low concentrations of dissolved salts [6].

1.2 Significant studies of lead removal

The advancement of today was seeded by research years ago, and progress of tomorrow will be the result of today's planned research. Comprehension of allied literature of past studies is very much indispensable for any investigation work to formulate sound methodology that acts as a guiding source during advancement of research. Chitin and chitosan are the frequently used biomaterials in mitigation of water pollutants [1, 8]. But, low mechanical resistance and high solubility in acid medium limit their applications in applied conditions, as observed in water treatment. However, such disadvantages can be overcome by preparation of chitin- and chitosan-based biocomposite matrix by doping techniques. The biopolymer-based adsorbent media can be chemically and mechanically stabilized by doping chitosan matrix to alter its solubility and brittleness. Herein, a novel de-fluoridation procedure has been developed, which successfully removed fluoride from water and exhibits greater competitive adsorption capacities than those of commercial activated alumina (even at broad range of pH). The novel aspects of research work were inferred from review of literature and found that this developed Pb^{2+} removal technique of water works by the sorption of fluoride onto bioadsorbents in efficient way in terms of magnitude of crisis, and this research is a small attempt for a big problem.

1.3 Lead properties

Lead is a soft, malleable heavy metal that belongs to the carbon group with symbol Pb (Latin: Plumbum) [1–4, 6]. It is also found in the earth's crust, and the proportion gets increased with time due to radioactive uranium disintegration. Atomic number of lead metal is 82, atomic weight 207.2, melting point 327.5°C, density 11.34 g/cm^3, and electronic configuration [Xe] $4f^{14}$ $5d^{10}$ $6s^2$ $6p^2$. Metallic lead has a bluish-white color after being freshly cut, but it soon tarnishes to a dull grayish color when exposed to air. Lead metal exists as the natural abundance in four stable isotopic forms namely Pb^{204}, Pb^{206}, Pb^{207}, and Pb^{208}. Metallic lead PbO rarely occurs in nature; nevertheless, the metal exists in two main oxidation states Pb^{2+} and Pb^{4+} with more stable divalent form in the aquatic environment [2]. Three lead oxides are known, that is, lead monoxide (PbO), lead dioxide (PbO_2), and lead tetroxide (Pb_3O_4) also called minium. Lead is found in ore with other heavy metals such as zinc, silver, and copper and is extracted together with these metals. The main lead mineral is galena (PbS) that contains 86.6% lead by weight. However, as a result of human activity in the atmosphere, lead metal exists mainly as cerussite ($PbCO_3$) and anglesite ($PbSO_4$) [5, 7, 10].

2. Environmental sources of lead

The presence of lead in water, air, and soil environment even in traces has detrimental effects on plants and animals. The natural sources of lead are soil erosion, volcanic eruptions, sea sprays, and bush fires. The anthropogenic activities dispersed lead compounds throughout the environment. Lead is transferred continuously through air, water, and soil by natural chemical and physical processes such as weathering, runoff, precipitation, dry deposition of dust, and stream/river flow [1–4, 6].

2.1 Air

The ingestion of lead through food and water is much greater than that of urban air. Inhaled lead of about 20–40%, which is derived from the air, is much more readily absorbed [6]. The air released from the combustion of fossil fuels, especially leaded gasoline, and several industrial processes dealing with storage batteries, mining, and ore smelting operations are of major concern regarding lead pollution in the atmosphere. Children are more susceptible to airborne lead poisoning, and lead absorbed in the guts of infants and young children is estimated to be 5–10 times greater than in adults [2, 11]. Faulty removal of lead-based paint, street dirt, and household dust absorbed through the lungs, skin, and intestinal tract is also responsible to airborne lead contamination. Cigarette smoke is also a significant source of lead exposure due to lead arsenate insecticides sprayed on tobacco. Most of lead ions reside in smoke-ash; however, studies have estimated 20 nanogram of lead per cigarette smoke, leading toxic effects on lungs of smoker.

2.2 Food

Food sources can be contaminated with lead due to spraying of lead arsenate insecticides or lead accumulations during food processing. Exclusively imported pottery such as ceramic cookware possess lead-containing glaze is a common source of lead toxicity. Lead solder used for sealing of food cans, especially the acidic foods such as tomato, okra and orange, grapefruit, or cranberry juices are also key sources of lead intake. Canned juices and canned baby foods such as evaporated milk may contain up to 100 and 200 μg of lead per liter, respectively [12], as high as 300 μg/day total lead uptake through food. Ingestion of peeled lead-based paint pica in children causes poisoning [2, 8, 13].

2.3 Soil

The toxic lead compounds are strongly adsorbed onto the upper layers of soil and do not leach into the subsoil. The average residence time of lead in the atmosphere is about 10 days. The presence of high concentration of lead in the soils results in lead contaminated fruits and vegetables. Particulate pollutants are emitted through leaded paints, leaded gasoline, and lead in pipes that can also contaminate the soil with heavy metal lead/Pb^{2+} [14].

2.4 Water

The solubility of lead compounds is the highest in soft and acidic water, and it is a function of pH, hardness, and salinity of water sample. Several industries engaged in releasing of industrial wastewater effluents, lead acid batteries, fertilizers, pesticides, and mining waste; metallurgical, chemical, and petrochemical industries are prominent sources of releasing toxic lead in the water stream. Lead rarely occurs in natural water bodies, but the major source of lead in drinking water is from lead-based plumbing materials. The corrosion of such leaded pipes/fixtures enhances lead amount in community water. Old homes constructed before 1986 owing lead pipes, fixtures, and solder are the main contributor to lead in tap water. Other water delivery systems such as lead solder used to join copper pipes, brass in faucets, coolers, and valves are liable for lead content in water. Older submerged pumps used in well water can also contain leaded-brass works. Seawater and river water contain 2–30 ppt and 3–30 ppb of lead content, respectively. Phytoplankton contains 5–10 ppm, freshwater fish 0.5–1000 ppb, and oyster 500 ppb of lead/Pb^{2+} concentration [15].

3. Control of lead contamination

In India, the Central Pollution Control Board (CPCB) has carried out a major ground water quality survey, and the report recognized about 20 critical sites of ground water pollution in various states of India. CPCB found that industrial effluents are the primary and major cause for ground water pollution [1–4, 6, 11]. The major heavy metal contamination sites including lead metal in Indian scenario have been reported in **Figure 1**. The chemical quality of ground water as monitored by CPCB India showed that the states such as Haryana, UP, Punjab, West Bengal, Tamilnadu, and Telangana own heavy contaminations of lead metals [12, 13].

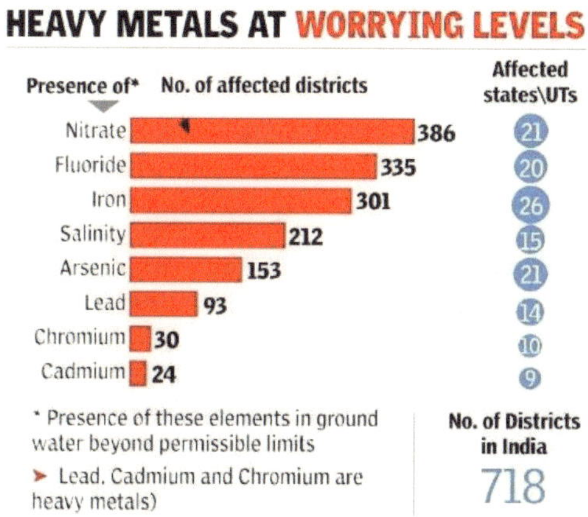

Figure 1.
Heavy metal contaminations in various Indian states. Sources: Timesofindia.indiatimes.com/articleshow/65204273.cms?utm_source=contentofinterest&utm_medium=text&utm_campaign=cppst.

3.1 Regulatory aspects of lead in water

According to the Indian Standard Institution (ISI), the tolerance limit for discharge of lead into drinking water is 0.05 mg/L and land surface water 0.1 mg/L [16]. The World Health Organization (WHO 1995) had proposed safe total lead limit of 50 ppb in drinking water, which was further decreased to 10 ppb [1–4, 6]. The permissible limit of lead ions in drinking water as set by the European Union (EU), the United States Environmental Protection Agency (USEPA), and Guidelines for Canadian Drinking Water Quality in 2012 is 10, 15, and 10 ppb, respectively. However, more recently, an EPA document recommends a zero lead/Pb^{2+} value in national primary drinking water standard [17].

4. Symptoms and health effects of lead

The human body contains around 120 mg of lead, and 10–20% of lead is absorbed by the intestines. The doorway of poisonous lead in human system mainly through contaminated air, food, and water sources manifests overt and detrimental health problems. Lead is a cumulative poison, and it elucidates destructive effects on almost every physiological systems namely musculoskeletal, nervous, cardiovascular, digestive, reproductive, excretory, endocrine, and metabolic system. Lead is highly toxic and carcinogenic even at low concentration [2, 6]. International Agency for Research on Cancer (IARC) classifies inorganic lead compounds as probably carcinogenic to humans (Group-2A). The National Toxicology Program (NTP-2005) of the US Department of Health and Human Services concluded that "Lead and lead compounds are reasonably anticipated to be human carcinogens" [16, 17].

4.1 Effects of lead on children

Children are usually more vulnerable for toxicity of lead due to immature blood-brain barrier and the fact that lead can easily cross blood brain and placental barrier, and thus they readily absorb a larger amount of lead per unit body weight than adults. The serious effects of Pb^{2+} on health of children include encephalopathy, peripheral neuropathy, cognitive impairment, and personality disorders (USEPA, 1986a). If the neuropathy is severe, the lesion gets permanent. Lead toxicity showed dark blue lead sulfide line at the gingival margin of the person. It is found that the fatal effects of lead (II) are marked by seizure, coma, and death if not treated immediately according to the USEPA studies [2, 8, 18]. Evidence suggests that lead may cause fatigue, irritability, information processing difficulties, memory problems, a reduction in sensory and motor reaction times, decision-making impairment, and lapses in concentration [12]. Lead interferes with heme biosynthesis by changing the activity of three enzymes δ-aminolevulinic acid synthetase (δ-ALAS), δ-aminolevulinic acid dehydratase (δ-ALAD), and ferrochelatase and thus affects the hematological system. The presence of Pb^{2+} ions above 70 mcg/dL in human blood exhibited microcytic and hypochromic anemia being characterized by hemoglobin reduction and basophilic stippling of erythrocytes along with a shortened life span of red blood cells (erythropoiesis) [6, 11, 12]. Increase blood lead level shows decreased intellectual capacity and IQ level of children by four to seven points for every 10 μg/dL [11]. Attention deficit hyperactivity disorder (ADHD) hearing impairment in child may disrupt peripheral nerve function (ATSDR 2007).

4.2 Effects of lead on adults

Lead toxicity affects renal system as it causes many effects such as aminoaciduria, glycosuria, and hyperphosphaturia, that is, Fanconi-like syndrome [2, 19]. Kidney disease, both acute and chronic nephropathy, is a characteristic of lead toxicity [12]. Lead poisoning inhibits excretion of the waste product urate that causes a tendency for gout, that is, saturnine gout. Occupationally, lead exposed individual tends to have more hypertension than normal people and augmented risk for cardiovascular diseases, myocardial infarction, and strokes [20]. Lead toxicity includes gastrointestinal disturbances – abdominal pain, cramps, constipation, anorexia, and weight loss – immune suppression, and slight liver impairment. In adults, high levels of lead can cause headaches and disorders of mood, thinking, memory, irritability, lethargy, malaise, and paresthesia. There is also some evidence that lead exposure may affect adult's postural balance and peripheral nerve function, which can cause tremors or weakness in fingers, wrists, or ankles [2, 21]. Lead poisoning affects the human male reproductive system by decreasing the sperm count and increasing the abnormal sperm frequencies. Women are more susceptible to lead poisoning than men, and lead toxicity causes menstrual disorder, infertility miscarriages, and stillbirths. Lead inhibits several enzymes required for the synthesis of heme, causing a decrease in blood hemoglobin. Lead interferes with a hormonal form of vitamin D, which affects multiple processes in the body, including cell maturation and skeletal growth. Lead poisoning is also known to cause psychotic behavior such as hyperactivity or schizophrenia.

5. Assorted treatment processes for lead removal

Detrimental heavy metals must be removed from the environment in particular water so as to protect the human beings and the environment [2–4, 6, 22]. To accomplish the increased stringent environmental regulations and maximum permissible limit of contaminant in water, a wide range of treatment technologies such as chemical precipitation, coagulation flocculation, flotation, ion exchange, membrane filtration, electrochemical treatment technologies, adsorption [5, 21], and bio-adsorption are most frequently examined for the mitigation of heavy metals from wastewater [2, 8]. Certain merits and demerits of various physicochemical methods used for the mitigation of Pb^{2+} from wastewater are mentioned in **Table 2**.

Entry	Treatment techniques	Merits	Demerits
1	Physicochemical precipitation	Low-cost, easy operation, high efficiency	Huge sludge/waste formations, addition burden of solid-waste disposal
2	Adsorption	Cheap, simple conditions for operation, wide pH workability, high binding capacity/efficiency	Less selective, create large wastes/secondary pollutants
3	Membrane filtration	Small space requirement, low pressure, high separation selectivity	Highly expensive, membrane fouling
4	Electrodialysis	Elevated separation selectivity, great efficiency	Costly, membrane fetid, needs more energy/power
5	Photo-catalysis	Remove metals, organics, green process, less byproducts	Limited utilities, prolong time

Table 2.
The merits and demerits of various physicochemical methods used for the mitigation of Pb^{+2} from wastewater [2, 22].

The reduction of Pb^{2+} in wastewater to 0.05–0.10 mg/L level is required before its discharge to outlet. Certain frequently used methods for remediation of Pb^{2+} from water are as follows:

a. **Precipitation:** It is usual practice use for remediation of Pb^{2+} of ppm level contamination from water. Pb^{2+} salts are insoluble in water and yet get entrapped as precipitates via treatment with certain chemicals such as soda lime, bisulphite, or ion exchangers in a practicable manner. This method uses many chemicals and ultimately generates huge solid wastes and thus poses a burden of sludge disposal. Besides the precipitation, techniques are nonspecific and noneffective for low concentration of Pb^{2+} ions [2].

b. **Ion exchange:** Ion exchange is another methodology use for the mitigation of Pb^{2+} ions from water. This is quite cheaper than the other known methods; besides ppb levels, Pb^{2+} ion removal can be achieved at large-scale workups. In fact, ion exchangers are natural materials namely certain clays, functionalized porous, gel polymer, zeolite, montmorillonite, and soil humus or some synthesized resins that hold positive/cation as well negative/anion exchanging parts with the other ions in solution owing to a better affinity from the surroundings [5, 7, 8, 10, 18, 23]. Certain artificial matrixes like organic resins are usual for ion exchangers used for the removal of Pb^{2+} ions from water/wastewater. However, such man-made ion exchangers are disadvantageous as nontolerant high level of metal due to matrix fouls [1, 5, 7, 8, 10, 18, 23]. Furthermore, synthetic/natural ion exchangers are non-selective and extremely susceptible to altered pH conditions. Ion-exchange methods have been widely used to remove heavy metals from water due to their many advantages, such as high treatment capacity, high removal efficiency, and fast kinetics [24]. The synthetic resins are most commonly preferred in ion exchange process as they effectively remove the heavy metals from the solution. The uptake of heavy metal ions by ion-exchange resins is rather affected by certain variables such as pH, temperature, initial metal concentration, and contact time. Charges present of ionic pollutant also controls ion-exchange phenomenon as reveled in purolite C-100 cation exchange resin carried abatement of Ce^{4+}, Fe^{3+}, and Pb^{2+} ions from aqueous solutions. Moreover, natural zeolites have also been widely used due to its low cost to remove heavy metal from aqueous solutions. Many researchers have demonstrated that zeolites exhibit good cation-exchange capacities for heavy metal ions under different experimental conditions [2–8, 10–12, 23, 25]. Clinoptilolite, natural zeolite, was extensively used to remove Pb^{2+} with an initial concentration of 2072 mg/L, at optimum pH 4, with an adsorption capacity of 0.21–1 meq/g in fixed bed and batch mode operation style. Clinoptilolite was studied for the removal of Pb^{2+} with an initial metal ion concentration of 1036 mg/L, at optimum pH 4, with a removal efficiency of 55% in batch mode experiments [26]. Inadequacies in use of this process are that ion exchange resins must be regenerated once exhausted, which in turn the regeneration eventually causes serious secondary pollution. Ion exchange process is not economical and cannot be used on large scale [5, 7, 8, 10, 18, 23–31].

c. **Electrowinning:** Electrowinning is also called as electroextraction being frequently used by mining and metallurgical operations for leaching and acid draining. Metal transformation industries employed electrowinning and electrodeposition for amputation and recovery of metal lead ions at insoluble anodes. In electrowinning, metals are electrodeposited from its ores via a leaching route. Electrorefining uses a similar process to remove impurities from a metal as similar to electrorefining used for eliminating impurities from metal. In-bulk conditions, electroplating occurs economically in straightforward purification of metals. In this process, a current is passed through an inert anode via a liquid leach containing lead metal and subsequently deposited at electroplate/cathode [2, 25].

d. **Electrocoagulation:** Electrocoagulation is an electrochemical process, which uses electrical charges to remove Pb^{2+} from water in an efficient manner, while contaminants are maintained in solution. In this technique after neutralization of Pb^{2+} ions in the solution, the residual coagulant aids destabilization and precipitation of other reverse charges/counter ions. Electrocoagulation is also performed in number of ways such as radio-frequency diathermy or short wave electrolysis. This electrocoagulation is capable to remove Pb^{2+} ions due to the fact that it is hard to take out by filtration or chemical treatments. Various electrocoagulation devices are known with complexity from plain anode and cathode to the larger and more complex device using manageable electrode potentials, passivation, anode consumption, cell redox potentials, ultrasonic, UV ray, and advanced oxidation processes.

e. **Cementation:** Cementation is a heterogeneous process similar to precipitation technique wherein metal ions like Pb^{2+} are reduced into zero valence at a solid metallic interfaces. This is more commonly used for refining leach solutions. In its solution, Pb^{2+} ions are precipitated in the presence of other solid via electrochemical system as metal with higher oxidation potential bypasses in solution and restore lower oxidation potential metals.

f. **Chemical precipitation:** Chemical precipitation is an effective and most widely used process in the industry due to its relative simplicity and inexpensive operation [1–4, 6]. In precipitation processes, chemicals react with heavy metal ions to form insoluble precipitates that can be separated from the water by sedimentation or filtration. Conventional precipitation technique consists of hydroxide precipitation and sulfide precipitation.

g. **Hydroxide precipitation:** The most widely used hydroxide precipitation technique is relatively simple and low cost with the ease of pH control. The pH range of 8–11 exhibits the decreased solubility of various metal hydroxides. The coagulants such as alum, iron salts, and organic polymers are used to enhance the removal of heavy metals from wastewater. Lime is the preferred choice of the base used in hydroxide precipitation at industrial settings. To increase lime precipitation, fly ash was used as a seed material. By using fly ash-lime carbonation treatment, the initial concentrations of 100 mg/L of Cd^{2+}, Cu^{2+}, Ni^{2+}, Zn^{2+}, and Pb^{2+} from synthetic effluents can be reduced to 0.08, 0.14, 0.03, and 0.45 mg/L, respectively, at optimum pH 7 with a removal efficiency of 99.37–99.6% [22, 32, 33]. Hydroxide precipitation has some limitations despite of its wide usage. First, it generates large volumes of relatively low-density sludge and consequently poses dewatering and disposal problems. Second, the metal hydroxide precipitation will be inhibited due to the presence of complexing agents in the wastewater, and third, as some metal hydroxides are amphoteric, accordingly in case of mixed metal pollution in water, ideal pH for one metal may put other metals back in solution.

h. **Sulfide precipitation:** Abatement of toxic heavy metal ions using sulfide precipitation is also an effective process. One of major advantages of sulfide precipitation over hydroxide precipitation is that sulfide precipitates are nonamphoteric and of lower solubility than hydroxide precipitates. Similarly, sulfide sludge demonstrated better thickening and dewatering characteristics than the corresponding sludge of metal hydroxide. The sulfide precipitation process can attain a higher extent of heavy metal removal over a broader pH range than hydroxide precipitation [1–3]. The technique of heavy metal removal by sulfide precipitation involves initial generation of H_2S gas at low $pH < 3$ (Eq. (1)) and subsequent adsorption at higher pH 3–6 (Eq. (2)) as shown below:

$$FeS(s) + 2H^{+}(aq) \rightarrow H_2S(g) + Fe^{2+}(aq) \tag{1}$$

$$M^{2+}(aq) + H_2S(g) \rightarrow MS(S) + 2H^{+}(aq) \tag{2}$$

The sulfide precipitation studies revealed that Cu^{2+}, Zn^{2+}, and Pb^{2+} with the initial concentration of 0.018, 1.34, and 2.3 mM get easily precipitated out at optimum pH 3 with a removal efficiency of 100, >94, and >92%, respectively [1–4, 6]. Investigation showed that pyrite and synthetic iron sulfide are used to remove/precipitate out Cu^{2+}, Cd^{2+}, and Pb^{2+} from water/wastewater [3, 4, 6]. The major drawback of use of sulphide precipitation is the evolution of toxic H_2S fumes in acidic conditions. Therefore, this precipitation needs to be carried out in neutral or basic medium. Furthermore, metal sulphide precipitation is likely to form colloidal precipitates that cause troubles in separation either by settling or by filtration processes. Chemical precipitation is appropriate to treat wastewater containing high concentration of heavy metal, and it is ineffective for low metal concentration. It is not economical and can produce large amount of sludge as a secondary pollutant [34].

i. **Membrane filtration:** Membrane filtration technologies used in heavy metal removal are easy in operation and highly efficient and space saving, but its tribulations such as high cost, complex operation, membrane pollution, and low permeate flux have restricted their use in heavy metal removal. The membrane filtration technique includes ultra-filtration, reverse osmosis, nanofiltration, and electrodialysis [1–8, 10, 23].

j. **Ultra-filtration:** To obtain high removal efficiency of metal ions, the micellar enhanced ultrafiltration (MEUF) and polymer enhanced ultrafiltration (PEUF) were suggested. MEUF was first introduced by Scamehorn et al. in the 1980s for the removal of dissolved organic compounds, and multivalent metal ions from aqueous solutions used MEUF to remove Cd^{2+}, Cu^{2+}, Ni^{2+}, Zn^{2+}, and Pb^{2+} from synthetic water using anionic surfactants [5, 7, 8, 10]. Ferella et al. studied MEUF to remove Pb^{2+} and AsO_4^-, with an initial concentration of 4.4 ppm to 7.6 mg/L at optimum pH 7.5 with a removal efficiency of 99 and 19% for Pb^{2+} and AsO_4^-, respectively [5, 7, 8, 10, 23]. The recovery and reuse of exhausted surfactant is necessary; otherwise, it enhances treatment cost and in addition causes secondary pollution due to solid waste/sludge disposal problem. Hence ultra-filtration technique has not attained wide applicability at industrial scale [12].

k. **Reverse osmosis:** Reverse osmosis (RO) is progressively more popular wastewater treatment alternative in chemical and environmental engineering. In reverse osmosis process, the water is to be purified and allowed to pass through the semi-permeable membrane and at the same time rejecting the contaminants. RO systems are yet to be broadly applied for wastewater treatment. Reverse osmosis alone is not applicable for complete recovery and reuse of fluids. Pretreatment methods namely media filtration, pH adjustment, and use of anti-precipitants are required prior to the reverse osmosis. The main disadvantage of RO is the high power utilization owing to the pumping pressures and the reinstallation of the semi-permeable membranes. Reverse osmosis and electrodialysis use semi-permeable membranes for the removal and revival of Pb^{2+} from water. In this technique, cation and anion membranes from water, which are tied to the electrodes in electrolytic cells under constant electrical supply and subsequent allied ions, get drifted. Treatment characteristics are optimized with respect to Pb^{2+} concentration in the effluents, pH, temperature, and flow volume. Reverse osmosis and electrodialysis (mentioned below) techniques use semi-permeable anionic/anionic membranes which possess certain drawbacks like high cost, generates huge sludge, low retention capacity, and less selectivity and large power usage. Thus, both these techniques are neglected while adsorption method is preferred for the remediation of heavy metal contaminants from water/wastewater [1–8, 10–12, 23].

l. **Electrodialysis:** Electrodialysis (ED) is another membrane technique used for the separation of ions through membranes from one solution to another under the influence of electric field. This process has been widely used for the treatment of brackish water, industrial effluents, recovery of useful materials from effluents, and salt production [3]. ED is also used as a potential method for wastewater treatment containing heavy metals. The effects of operating parameters on Pb^{2+} mitigation from water have been investigated by ED, and the results revealed that Pb^{2+} separation improved with increasing voltage and temperature [12].

m. **Coagulation and flocculation:** Coagulation followed by sedimentation and filtration is one of the most significant methods for wastewater treatment, but coagulation method is restricted only to the hydrophobic colloids and suspended particles. Many coagulants such as aluminum, ferrous sulfate, and ferric chloride are widely used for the effective removal of wastewater pollutants. Flocculation is the process by which fine particulates are caused to clump or agglomerate together into flocs. In general, PAC (polyaluminum chloride), polyferric sulfate (PFS) and polyacrylamide (PAM) are widely used flocculants in wastewater treatment, but practically, they are ineffective in heavy metal removal. The flocculation by humic acid (HA) is also studied. It is also studied to reveal better removal of Pb^{2+} and Zn^{2+} from solution by binding such ions to HA and then coagulating-flocculating with cationic polyelectrolyte polydiallyl dimethyl ammonium chloride (poly-DADMAC). Generally, coagulation-flocculation cannot treat the heavy metal wastewater completely [24]. So, it must be combined with other treatment techniques. Besides, coagulation-flocculation involves chemical utilization and sludge formation as a secondary pollution.

n. **Floatation:** Recently, flotation has found widespread use in heavy metal removal from wastewater due to its several advantages such as high metal selectivity, high removal efficiency, high overflow rates, low operating cost, and more concentrated sludge production [27]. Flotation has been implemented to remove heavy metal from a liquid phase using small air bubble attachment. The main flotation processes are dissolved air flotation (DAF), ion flotation, and precipitation flotation used for metal ion abatement from water. Among these techniques, ion flotation has been proved to be a promising method for the wastewater treatment. The ion flotation is an adsorptive bubble separation techniques based on imparting the ionic metal species hydrophobic by the use of surfactants and removal of these hydrophobic species from wastewater by the passage of air bubbles reported the 100% removal of copper (II) and Pb^{2+} from water at pH 6 and 7, respectively, using ion floatation technique. Despite of several advantages, the shortcoming in use of flotation techniques involves high initial capital cost, high maintenance, and expensive operation [12, 27].

o. **Electrochemical treatment:** Electrochemical wastewater treatment technologies are rapid, well controlled, and capable of producing less sludge. It requires few chemicals and possesses better metal reduction capacity. Electrochemical technologies include electrocoagulation, electroflotation, and electrodeposition. Studies of electroflotation technique revealed more stress on separation of heavy metal ions such as iron, nickel, copper, zinc, lead, and cadmium with maximum removal [1–4, 6, 11]. The quantitative electrodeposition of copper and lead ions onto specially designed palm shell AC electrodes has been reported in the recent investigation studies [1–4, 6]. Nevertheless, the requirement of high initial capital cost and the expensive electricity supply restricts wide usage and application of electrochemical techniques in wastewater treatment [6, 23].

p. **Adsorption:** Adsorption is defined as the process where a solute is removed from the liquid phase through contact with a solid adsorbent, which has a particular affinity for that particular solute [1–4, 6, 10– 12]. All adsorption processes mainly depend on the equilibrium and mass transfer rates, which involve the accretion of substances at the interface of liquid-liquid, gas-liquid, gas-solid, or liquid-solid interface. The substance being adsorbed is adsorbate, and adsorbing material is adsorbent. The reverse phenomenon of adsorption is termed as desorption where a solute/pollutant gets released from/through an adsorbent. The process of adsorption can be classified as physical adsorption/physisorption and chemical adsorption/chemisorption, which depends on the forces involved in adsorbate-adsorbent interactions. Adsorption is now recognized as an effective and economic method for the mitigation of wide variety of water pollutant by sorption. Among the aforementioned technologies, adsorption has been preferred due to its flexible operation and capabilities to generate high quality treated effluent as well as adsorbent by desorption. Several adsorbents such as agricultural wastes, carbon nanotubes, biosorbents, industrial byproducts, natural substances, and activated carbon have been studied for the heavy metal wastewater treatment. Numerous studies for the removal of heavy metal ions from water onto activated carbon are well known due to its high surface area and large micropore/mesopore volumes [1–4, 6, 11, 12, 24, 27–29, 34]. The production of activated carbons from abundantly available agricultural wastes converts unwanted, additional agricultural waste to useful valuable adsorbents. Adsorption capacity of natural materials for metal Pb^{2+} removal from water is shown in **Table 3**.

The activated carbon of agricultural wastes such as coconut coir pith, nutshells, oil palm waste, rice husk, and peanuts is used for the removal of lead metal from water [1–4, 6, 11, 12]. *Eucalyptus camaldulensis* Dehn. bark agricultural waste bagasse was studied for the removal of contaminants from water. Carbon nanotubes are relatively new adsorbents, which show great potential for the removal of heavy metal like lead ions from water. The adsorption of Pb^{2+} ions onto acidified multi-walled carbon nanotubes (MWCNTs) showed the removal capacity of 85 mg/g of adsorbent at NTP conditions [1–4, 6, 11, 12]. The carbon nanotubes (CNTs) that are used as effective adsorbents for Pb^{2+} adsorption at pH 5, contact time 80 min, and agitation speed 50 rpm were found to have maximum adsorption capacity of 102.04 mg/g. Agricultural waste materials that act as a potential adsorbent for sequestering heavy metal ions from aqueous solution were also investigated [5, 7, 8, 10, 23]. In addition, chemically modified plant wastes that act as low cost adsorbents for heavy metal uptake from contaminated water were also reported [25]. Finally, industrial byproducts such as lignin, diatomite, clino-pyrrhotite, aragonite shells, clay, and peat were used to remove Pb^{2+} from water [1, 25, 34]. However, the adsorption efficiency of such industrial wastes is pH dependent and better works only in the pH range from 2 to 6 (mostly acidic), while competitive sorption capacity falls relatively in multicomponent/metal systems.

Entry	Assorted adsorbents	Pb^{2+} adsorption capacity (mg/g)
1	Zeolite, clinoptilolite	1.6
2	Modified zeolite/MMZ	123
3	Clay (HCl-treated)	81.02
4	Poly(methoxyethyl)acrylamide-doped Clay	85.6
5	Calcinated phosphates	155
6	Activated phosphate	4.0
7	Zirconium phosphate	398
8	Almond shells	8.0
9	Palm shell oil	3.4
10	Maize cope and husk	456
11	Ecklonia maxima – marine alga	235
12	*Oedogonium* species	145
13	*Nostoc* species	93.5
14	*Bacillus* – bacterial biomass	467

Table 3.
Adsorption capacity of natural material for Pb^{2+} removal from water [24].

6. Selective adsorbents for lead remediation

Chitosan has an excellent adsorption capacity due to its low chemical and mechanical strength [1–4, 6, 11, 12]; its fullest adsorptive capacities are not met [5, 7, 8, 10]. Thus, several attempts of chitosan modification were carried out in previous research studies. In present research work, the materials namely graphite, iodate, and activated carbon of *Luffa cylindrica* are used for the modification of chitosan by blending/doping or impregnation method. The resultant adsorbents obtained are called graphite-doped chitosan composite (GDCC), iodate-doped chitosan composite (IDCC), and activated carbon of *Luffa cylindrica*-doped chitosan biocomposite (ACLFCS) [5, 7, 10]. The graphite has been selected for the study so as to check the adsorptive potential of chitosan-graphite composite and its prospective applications in wastewater treatment. The deacetylated amino groups in chitosan can be chemically modified easily [5, 7, 8, 10, 23]. Iodine molecule has a characteristic property of donating a new function to the host material. The iodate salt has been selected for modification of chitosan so that to find out whether the new surface chemistry of iodate-doped chitosan composite has an impact on Pb^{2+} removal from water or not. *Luffa cylindrica* is mainly a lignocellulosic material composed of cellulose, hemicelluloses, and lignin (60, 30, and 10% by weight, respectively) belonging to the Cucurbitaceae family.

Agricultural residues are abundant in a number of developing countries such as India, Korea, China, Central America, and Japan. Consequently, these disposed, unconventional, and widely available *Luffa cylindrica* fibers can be transformed into an activated carbon, which is a carbonaceous material that possesses highly developed porosity, a large surface area, relatively high mechanical strength, and a variety of functional groups on its surface [10, 34]. The transformation of agricultural residue into an activated carbon ultimately provides a way to reduce its environmental burden or hazards. The cationic nature of chitosan and the anionic nature of activated carbon of *Luffa cylindrica* can produce a stable, granular biocomposite due to two oppositely charged interactions. The purpose of doping of chitosan with *Luffa cylindrica* activated carbon is to explore the expected synergistic effects achieved through the incorporation of certain functionalities in the resultant biocomposite that could be responsible for the adsorption of Pb^{2+}. The choice of adsorbents for Pb^{2+} removal from water is concerned with its high adsorption efficiency, safety and simplicity for use, ease for maintaining, minimal production of residual mass, low capital cost, and nontoxicity. The synthesized adsorbents such as GDCC, IDCC, and ACLFCS satisfy all such requirements [24, 34]. To the best of our knowledge, there were no published reports on the removal of Pb^{2+} ions from water using GDCC, IDCC, and ACLFCS. In fact, the ability of above synthesized adsorbents to adsorb other heavy metal ions was also not reported. This provided a way for more adsorption studies for Pb^{2+} ions to be conducted by using these synthesized adsorbents. Batch adsorption of Pb^{2+} is studied as a function of various parameters, namely pH, agitation time, adsorbent doses, and initial Pb^{2+} concentration onto specially developed adsorbents like GDCC, IDCC, ACLFCS, raw chitosan, and activated carbon of *Luffa cylindrical* plant [1–4, 6]. Adsorption kinetics and isothermal studies are conducted to know the mechanism of batch adsorption of Pb^{2+} removal from water [1–4, 6, 10].

7. Case studies for Pb^{2+} removal

Chitosan is a natural nitrogenous amino polysaccharide found to be highly selective for the uptake of toxic heavy metal ions from contaminated water resources

[5, 7, 8, 10, 23, 24, 34]. Due to its adsorptive potentiality, it has been popularly used for environmental cleanup or for the mitigation of toxic heavy metal ions from water/wastewater. Despite of its variety of advantages, it possesses some major inadequacies that impose an urge for its modification so as to allow its better fit for the wastewater treatment applications. The present research work dealt with the modification of chitosan by various techniques such as blending, impregnating, and doping it with inorganic/organic materials, and the resultant biocomposites can be effectively utilized for the mitigation of toxic Pb^{2+} from water. This case study used synthesized chitosan-doped composites which were characterizations through advanced techniques like X-ray diffraction (XRD), scanning electron microscopy (SEM), thermal gravimetric analysis (TGA), differential scanning calorimetry (DSC), Fourier transform infra-red spectroscopy (FTIR), elemental analysis, BET, and BJH pore size distribution for better understanding of adsorption mechanism.

The batch mode Pb^{2+} adsorption was studied as function of different operating parameters such as pH, adsorbent doses, initial Pb^{2+} concentration, and agitation/contact time. The adsorption isothermal studies were conducted by means of Langmuir and Freundlich models.

Adsorption kinetic studies conducted so as to know the time-dependent adsorption behavior of doped bioadsorbents toward the removal of Pb^{2+} from water. The adsorption kinetics revealed the existence of both heterogeneous surface and monolayer coverage of adsorbed Pb^{2+} ions via the pseudo-second-order kinetic process. Amid all biocomposites, iodate-doped composite bioadsorbent has been proved to be an effective adsorbent for the sorption of Pb^{2+} from water [5, 7, 8, 10, 33, 35]. An effective adsorbent dose of iodate-doped bioadsorbent was found to be 0.5 g/L with 99% efficiency achieved in 4 h [18].

The regeneration/desorption studies were done by desorption of exhausted adsorbents by means of several acid/alkali treatments, which provide the information about the respective utility of adsorbents in multifold cycles for batch mode adsorption of Pb^{2+} from water. Various studies demonstrated that the batch adsorption mode is used for the mitigation of Pb^{2+} from water using synthetically modified bioadsorbents namely raw chitosan (CS), GDCC, IDCC, ACLF, and ACLFCS [1–3, 7, 8, 18, 23, 29–31]. Chitosan-modified composites/biocomposites namely GDCC, IDCC, and ACLFCS were synthesized by means of impregnating/doping methodology [1–3, 7, 8, 18, 23, 24, 35]. Characterization studies of all synthesized chitosan composites/biocomposites were carried out by various instrumental techniques such as FTIR, TGA/DSC, XRD, BET surface area, BJH pore size distribution, SEM, and CHNS analysis. FTIR analysis of raw chitosan revealed that the absorption peaks are related to various functionalities namely –OH group, $-NH_2$ bending in amide, and C=O stretch in amide, and these functional groups were also responsible for Pb^{2+} ion adsorption onto the raw chitosan. Functional groups –OH, C=O, and $-NH_2$ were concerned with adsorption performance of Pb^{2+} onto GDCC [18, 24, 34].

Similarly, the functional groups –OH, C=O, C–N, iodate, and $-NH_2$ were responsible for Pb^{2+} adsorption onto IDCC. FTIR analysis of ACLF revealed that the absorption peaks are related to various functional groups namely –OH group, aromatic C–H stretch, aromatic C=C stretch, and phenolic C–O stretch [27–29]. The responsible and accurate sorption sites for Pb^{2+} adsorption onto synthesized adsorbents were ascertained by changes in FTIR band frequencies of assorted functional groups like –OH and –C=O, NH2, and – NH(C=O)CH3.

TGA of raw chitosan showed two steps of degradation, initial with 5% weight loss and the second with 46.28% weight loss. At the end of 955°C, total weight loss of raw chitosan was 70%. TGA/DSC analysis of pure graphite showed high thermal stability and displayed only 2.5% weight loss at the end of temperature 955°C. TGA of GDCC composite exhibited two steps of degradation, and at the end of 954.9°C, the total weight loss was found to be 35%. Thus, with respect to the raw chitosan, the GDCC showed more thermal stability. TGA/DSC analysis of IDCC indicated that this composite is less stable than raw chitosan [29, 30].

Thermogravimetric analysis (TGA) of ACLF revealed the high thermal stability where at the end of 910°C, it exhibited total weight loss of only 25%. XRD analysis of raw chitosan showed the broad diffraction peak at $2\theta = 20°$. XRD pattern of GDCC exhibited that the peaks of both raw CS and powdered graphite indicated the formation of single-phase composite. The peaks were obtained at $2\theta = 26.5°$ corresponding to graphite, and the broad peak at around $2\theta = 20°$ that was due to the chitosan decreased in intensity in GDCC composite, which confirms that graphite is doped onto the surface of raw chitosan [7, 8, 18, 23].

In IDCC, the XRD pattern is slightly broaden and showed a small hump at $2\theta = 25°$ and $2\theta = 34°$. The XRD pattern of IDCC after Pb^{2+} adsorption showed some additional crystalline peaks at $2\theta = 30°$ corresponding to Pb^{2+} ion adsorption. XRD analysis of ACLF showed the amorphous structure with two broad diffraction peaks at $2\theta = 24°$ and $2\theta = 42°$, which were the characteristic peaks of activated carbon. In ACLFCS biocomposite, the XRD pattern shows both the peaks of raw CS corresponding to $2\theta = 19.28°$ and a small hump at $2\theta = 42°$, which was a characteristic peak of ACLF [2]. The result indicates the successful doping and incorporation of ACLF with raw CS. SEM morphology of raw chitosan depicts uneven texture with bumpiness and porous cavities. SEM images of GDCC revealed flaky, smooth, shiny, and porous morphology with small amount of voids or cavities. SEM morphology of IDCC revealed that the bumpiness corresponding to raw chitosan was lost after doping it with iodate [2, 4, 7]. Similarly, in IDCC, the adsorbent surface was highly irregular and porous in nature with large number of round or elliptical shape cavities [7, 8, 18, 23]. SEM morphology of ACLF depicts the porous texture. ACLFCS possesses porous texture with round and elliptical shaped voids that can provide the adsorption sites to the adsorbate. SEM images of lead loaded adsorbent namely raw chitosan (CS), GDCC, IDCC, ACLF, and ACLFCS showed complete change of surface morphology [5, 7, 8, 10, 18, 23, 24, 34].

SEM image of lead loaded raw chitosan, IDCC, ACLF, and ACLFCS revealed surface morphology where the porous structure is quite shallow and not clearly seen.

The adsorption of Pb^{2+} by GDCC adsorbent showed deposition of whitish, sharp, needle-shaped crystalline mass observed onto its proactive surfaces [24]. The BET surface area of raw chitosan (CS), GDCC, IDCC, ACLF, and ACLFCS was found to be 9.923, 3.8, 6.984, 321, and 132 m^2/g, respectively [1– 4, 6–8, 18, 23]. The further lower in surface area in IDCC and GDCC with respect to raw chitosan and ACLFCS with respect to ACLF is due to doping of iodate and activated carbon of *Luffa cylindrica,* respectively, with raw chitosan structure. BET surface area of doped chitosan composites namely GDCC, IDCC, and ACLFCS was decreased due to blockage of internal porosities of adsorbents [2, 7, 8, 18, 23]. Adsorptive capacity of adsorbent for the removal of Pb^{2+} from water found increasing with rising surface area as viable for pure physisorption phenomenon [1–3, 5, 7, 8, 10, 18, 23, 25].

The physisorption was the limited phenomenon, while the chemisorption mechanism was mainly observed for all the synthesized bioadsorbents used for Pb^{2+} removal from water [2, 5, 7, 8, 18, 23]. The batch adsorption of Pb^{2+} as a function of pH, doses of adsorbent, initial Pb^{2+} concentration, and contact time has been carried out by using raw CS, GDCC, IDCC, ACLF, and ACLFCS bioadsorbents [1–8, 18, 23]. Maximum adsorption capacity of raw chitosan was found to be at an optimum parameter of pH 6, with 35 mg/L Pb^{2+} ion concentration at an adsorbent dose of

0.9 g/L achieved in contact time of 140 min. By using GDCC adsorbent, the maximum 98% removal of Pb^{2+} was observed at optimum pH 6. Results showed that the maximum adsorbent capacity was at a dose of 1 g/L and equilibrium time achieved at 120 min of contact time for 35 mg/L Pb^{2+} ion concentration. The batch adsorption studies for Pb^{2+} removal from water by using IDCC revealed maximum removal at optimum pH 6 with an adsorbent dose of 0.5 g/L in contact time of 240 min. The regeneration ability of IDCC adsorbent demonstrated 25–30% decreased percentage recovery of Pb^{2+} ions at the end of fourth adsorption-desorption cycle. The maximum adsorption of Pb^{2+} onto ACLF was exhibited at an optimum parameter of pH 5.5, with an adsorbent dose of 0.3 g/L achieved in 12 min of contact time for 35 mg/L Pb^{2+} concentration [2, 7, 8, 18, 23]. Batch adsorption of Pb^{2+} using ACLFCS biocomposite was pH dependent, and maximum 98–99% Pb^{2+} removal occurred at pH 5 in contact time of 15 min and at an optimum dose of 0.1 g/L [2].

The percentage recoveries of metal ions decreased by 28% at the end of fifth adsorption-desorption cycle due to saturation of adsorbent-binding sites [2, 8, 24]. Equilibrium adsorption experiments for all the adsorbents were studied at room temperature and data obtained fitted to Langmuir and Freundlich adsorption isotherm [1–8, 10, 18, 23, 29–31]. The maximum adsorption capacity for raw CS, GDCC, IDCC, ACLF, and ACLFCS was obtainesd as 8.77, 6.711, 22.22, 101.01, and 111.11 mg/g, respectively [2, 7, 8, 18, 23, 24]. The adsorption kinetics was analyzed using pseudo-first-order, pseudo-second-order, and intraparticle diffusion models [7, 8, 18, 23]. The adsorption experimental data better fitted with pseudo-second-order kinetic model for all the adsorbents namely raw CS, GDCC, IDCC, ACLF, and ACLFCS [2, 7, 8, 18, 23, 24]. All these synthetically modified chitosan-doped composite/biocomposite has been proved to be effective and economic adsorbents for the adsorption of Pb^{2+} from water.

8. Conclusions

Various methodologies are being developed for the mitigation of Pb^{2+} from water/wastewater including chemical precipitation, electrochemical reduction, ion exchange, reverse osmosis, membrane separation, and adsorption. All such approaches being developed must be cheaper and more effective with the concern of reducing further/secondary waste generations besides boosting the quality of treatments. Physicochemical adsorption is the preferred treatments seeking for cheap, biocompatible adsorbents get intensified nowadays. A number of techniques have been chosen for the treatment of Pb^{2+} polluted water/wastewater; however, the selected one must focus on certain crucial parameters such as pH, initial Pb^{2+}ion concentration, efficiency, and overall output as compared to other existing technologies besides socioenvironmental impacts and economic parameters such as fund investment and functioning costs. Last but not the least, its technical applicability, plant simplicity, and low cost are also guiding factors that played vital roles in the selection of the most suitable treatments. All such factors are most effective for opting suitable treatment techniques so as to protect the nature/environment.

9. Futuristic research

Following recommendations are made for the future scope:

- the modification of synthesized chitosan-based biocomposite should be investigated by using other techniques such as sol gel and hydrothermal;

- synthesized chitosan-based bioadsorbents are investigated for the removal of other toxic heavy metal ions namely cobalt, mercury, zinc, copper, bismuth, cadmium, chromium, and arsenic;
- bioadsorbents can also be investigated for the adsorption of Pb^{2+} contaminated waters;
- the chitosan-based biocomposite modification can be investigated in order to increase its surface area for effective adsorption of pollutants;
- the competitive removal of Pb^{2+} from water in its combined presence with other toxic heavy metal ions can also be studied;
- batch mitigation of Pb^{2+} onto the synthesized adsorbents at various temperatures may be conducted so as to determine its thermodynamic parameters;
- shelf life of synthesized and regenerated adsorbents also anticipates for investigation; and
- column mode of Pb^{2+} removal from industrial wastewater as futuristic step.

Acknowledgements

The author would like to thank the Head, Department of Chemistry, R.T.M. Nagpur University, Nagpur for providing lab facilities to carry out this research study and the Vice Chancellor, Nagpur University, Nagpur for approval and sanction of a research project under University Research Project Scheme, No. Dev./RTMNURP/AH/1672 (9), on dated September 24, 2016.

Author details

Rajendra Sukhadeorao Dongre
Department of Chemistry, RTM Nagpur University, Nagpur, MS, India

*Address all correspondence to: rsdongre@hotmail.com

References

[1] Dongre RS. Biological Activities & Application of Marine Polysaccharides. Croatia: IntechOpen; 2017. pp. 181-206. ISBN: 978-953-51-2860-1. DOI: 10.5772/65786

[2] Gedam AH, Narnaware PK, Kinhikar V. Blended composites of chitosan: Adsorption profile for mitigation of toxic Pb(II) ions from water. In: Chitin-Chitosan Myriad Functionalities in Science and Technology. ISBN: 978-1-78923-407-7. e-Book (PDF) ISBN: 978-1-83881-519-6. DOI: 10.5772/intechopen.74790

[3] Dongre RS. Chitosan-derived synthetic ion exchangers: Characteristics and applications. In: New Trends in Ion Exchange Studies. IntechOpen; 2018. pp. 21-42. ISBN: 978-1-78984-248-7. DOI: 10.5772/intechopen.78964

[4] Dongre RS. Rationally fabricated nanomaterials for desalination & water purification. In: Novel Nanomaterials. ISBN: 978-1-78923-089-5. e-Book ISBN: 978-1-83881-460-1. DOI: 10.5772/intechopen.74738

[5] Baghel A, Singh B. Emerging potable water technologies: Review paper. Defence Life Science Journal. 2016;**01**(2):113-126. DOI: 10.14429/dlsj.1.10739

[6] Dongre RS. Chitosan formulations: Chemistry, characteristics and contextual adsorption in unambiguous modernization of S&T. In: Hysteresis of Composites. ISBN: 978-1-78984-810-6. e-Book ISBN: 978-1-78984-811-3. DOI: 10.5772/intechopen.83391

[7] Dongre RS. Reinforced fabricated nano-composite matrixes for modernization of S&T in new millennium. Composite and Nanocomposite Materials—From Knowledge to Industrial Applications. IntechOpen: Croatia; 2020. DOI: 10.5772/intechopen.91305

[8] Dongre RS, Gedam A. Comparative study of sponge gourd derived biochar and activated carbon for bio-sorption and desorption of Pb(II) ions. Materials Today: Proceedings. 2019;**18**(1):887-900. DOI: 10.1016/j.matpr.2019.06.521

[9] Adhikari N, Sinha N, Narayan R, et al. Lead-induced cell death in testes of young rats. Journal of Applied Toxicology. 2001;**21**:275-277

[10] Zhang M. China: War without rules. Bulletin of the Atomic Scientists (SAGE0 Journal). 1999;**55**(6):16-18

[11] Abadin HG, Hibbs BF, Pohl HR. Breast-feeding exposure of infants to cadmium, lead, and mercury: A public health viewpoint. Toxicology and Industrial Health. 1997;**15**(4):1-24

[12] Abadin HG, Wheeler JS, Jones DE. A framework to guide public health assessment decisions at lead sites. Journal of Clean Technology, Environmental Toxicology and Occupational Medicine. 1997;**6**:225-237

[13] Adinolfi M. The development of the human blood-CSF-brain barrier. Developmental Medicine and Child Neurology. 1985;**27**:532-537

[14] Abbate C, Buceti R, Munao F, et al. Neurotoxicity induced by lead levels: An electrophysiological study. International Archives of Occupational and Environmental Health. 1995, 1986;**66**:389-392

[15] Adebonojo FO. Hematologic status of urban black children in Philadelphia: Emphasis on the frequency of anemia & elevated blood lead levels. La Clinica Pediatrica. 1974;**13**:874-888

[16] Agency for Toxic Substances and Disease Registry. Decision guide

for identifying substance specific data needs related to toxicological profiles; notice. Federal Register. 1989;**54**(174):37618-37634

[17] Agency for Toxic Substances and Disease Registry. Public Health Statement for Lead. Atlanta, GA: U.S. Department of Health and Human Services, Public Health Service; 1997

[18] Gedam AH, Dongre RS. Adsorption characterization of Pb(II) ions onto iodate doped chitosan composite: Equilibrium & kinetic studies. RSC Advances. 2015;5(67):54188-54201. DOI: 10.1039/C5RA09899H

[19] Kiani G, Soltanzadeh M. High capacity removal of silver (I), lead (II) ions by modified polyacrylonitrile from aqueous solutions. Desalination and Water Treatment. 2014;**52**(16-18):3206-3218

[20] Ghasemi M, Naushad M, Ghasemi N, Khosravi-fard Y. A novel agricultural waste based adsorbent for the removal of Pb(II) from aqueous solution: kinetics, equilibrium and thermodynamic studies. Journal of Industrial and Engineering Chemistry. 2014;**20**(2):454-461

[21] Acharya J, Kumar U, Meikap B. Thermodynamic characterization of adsorption of lead (II) ions on activated carbon developed from tamarind wood from aqueous solution. South African Journal of Chemical Engineering. 2013;**18**(1):70-76

[22] Abdel-Halim SH, Shehata AM, El-Shahat MF. Removal of lead ions from industrial waste water by different types of natural materials. Water Research. 2003;**37**(7):1678-1683

[23] Dongre RS. Nanomaterials via reconfiguration of skeletal matrix, Chapter 8. In: Nanostructures. Croatia: IntechOpen; 2019. p. 168. DOI: 10.5772/intechopen.86818. ISBN: 978-1-78923-994-2 1

[24] Gedam AH, Dongre RS, Bansiwal AK. Synthesis and characterization of graphite doped chitosan composite for batch adsorption of lead (II) ions from aqueous solution. Advanced Materials Letters. 2015;**6**(1):59-67

[25] Liu Z-R, Zhou L-M, Wei P, Lan H-H. Competitive adsorption of heavy metal ions on peat. Journal of China University of Mining & Technology. 2008;**18**(2):255-260. DOI: 10.1016/S1006-1266(08)60054-1

[26] Wanga S, Peng Y. Natural zeolites as effective adsorbents in water and wastewater treatment. Chemical Engineering Journal. 2010;**156**:11-24. DOI: 10.1016/j.cej.2009.10.029

[27] Dong L, Zhu Z, Qiu Y, Zhao J. Removal of lead from aqueous solution by hydroxyapatite/magnetite composite adsorbent. Chemical Engineering Journal. 2010;**165**(3):827-834

[28] Vilar VJ, Botelho CM, Boaventura RA. Influence of pH, ionic strength and temperature on lead biosorption by gelidium & agar extraction algal waste. Process Biochemistry. 2005;**40**(10):3267-3275

[29] Singanan M, Abebaw A, Vinodhini S. Removal of lead ions from industrial waste water by using biomaterials: A novel method. Bulletin of the Chemical Society of Ethiopia. 2005;**19**(2):289-294

[30] Singh D, Tiwari A, Gupta R. Phytoremediation of lead from wastewater using aquatic plants. Journal of Agricultural Technology. 2012;**8**(1):1-11

[31] Arbabi M, Hemati S, Amiri M. Removal of lead ions from industrial wastewater: A review of removal methods. International Journal of Epidemiologic Research. 2015;**2**(2):105-109

[32] Mondal MK. Removal of Pb(II) ions from aqueous solution using activated tea waste: Adsorption on a fixed-bed column. Journal of Environmental Management. 2009;**90**(11):3266-3271

[33] Singha B, Das SK. Removal of Pb(II) ions from aqueous solution and industrial effluent using natural biosorbents. Environmental Science and Pollution Research. 2012;**19**(6):2212-2226

[34] Hu J, Zhao D, Wang X. Removal of Pb(II) & Cu(II) from aqueous solution using multiwalled carbon nanotubes/iron oxide magnetic composites. Water Science and Technology. 2011;**63**(5):917-923. DOI: 10.2166/wst.2011.270

[35] Gedam AH, Dongre RS. Activated carbon from *Luffa cylindrica* doped chitosan for mitigation of lead (ii) from an aqueous solution. RSC Advances. 2016;**6**(27):22639-22652

Section 2

Lead Oxide Nanoparticles

Chapter 4

Synthesis, Characterization, Applications, and Toxicity of Lead Oxide Nanoparticles

Amra Bratovcic

Abstract

Over the past few years, the interest of material scientists for metal and metal oxide nanoparticles (NPs) is increasing dramatically because of their unique physicochemical characteristics such as catalytic activity and optical, electronic, antibacterial, and magnetic properties which depend on their size, shape, and chemical surroundings. Recently, several new routes of synthesis of lead monoxide (PbO) nanoparticles have been used, such as chemical synthesis, calcination, sol-gel pyrolysis, anodic oxidation, solvothermal method, thermal decomposition, chemical deposition, laser ablation, and green methods. Essentially, for the structural characterization of lead oxide nanoparticles, several spectroscopic, microscopic, and thermogravimetric methods of analysis are used. Lead oxide has been widely utilized in batteries, gas sensors, pigments, ceramics, and glass industry. Furthermore, lead oxide nanoparticles are graded as toxic and dangerous for the human health and environment. Therefore, there is an urgent need to develop new approaches and standardized test procedures to study the potential hazardous effect of nanoparticles on the human health and environment. The aim of this chapter is to provide an overview of the recent trends in synthesis of lead oxide nanoparticles, their characterization, possible applications, and toxicity.

Keywords: lead oxide nanoparticles, synthesis, characterization, applications, toxicity

1. Introduction

Today nanotechnology has become a top research field in the world. Nanotechnology is an interdisciplinary study which allows us to develop materials with new, interesting, and useful properties [1]. Nobel laureate Richard P. Feynman gave a lecture in 1959 at the annual American Physical Society meeting under title "There's plenty of room at the bottom," where he considered the possibility of direct manipulation of individual atoms as a more powerful form of synthetic chemistry than those used at that time [2]. Nanotechnology produced materials of various types at nanoscale level. Nanoparticles (NPs) are a large class of materials that include particulate substances having one dimension less than 100 nm at least [3]. Nanoparticles possess different properties and behave differently to the classical, larger building blocks of substances.

 IntechOpen

The world of nanotechnology is not new to mankind. The history of noble metal colloids can be traced back to ancient times. Cosmetic uses of lead-based chemistry were initiated in ancient Egypt about 4000 years ago [4].

Figure 1 illustrates an increase in the number of published papers on ScienceDirect by searching the keyword "PbO nanoparticles" on November 28, 2019, indicating that researchers' interest toward PbO nanoparticles continually increases since 2010.

Over the past few years, the interest of material scientists for metal nanoparticles has been booming because of its unique physicochemical characteristics such as a high specific surface area and a high fraction of surface atoms. The special properties of metal and metal oxide nanoparticles such as catalytic activity and optical, electronic, antibacterial, and magnetic properties [5–8] depend on their size, shape, and chemical surroundings [9] which can be designed by controlling the dimensions of these building blocks and their assembly via physical, chemical, or biological methods [10]. In particular, lead oxide electronic properties, i.e., bandgap and hence color, depend largely on the lead to oxygen ratios and crystal structures of various polymorphs.

Lead has a lot of oxide forms including PbO, PbO_2 (α, β, and amorphous), Pb_2O_3, and Pb_3O_4. Among them, lead monoxide (PbO) has been studied the most. Lead oxide is known as an important industrial material, which has been widely utilized in batteries [11], gas sensors [12], pigments and paints [13], ceramics [14], and glass industry [15] and as a catalyst in synthetic organic chemistry [16].

Different morphologies of lead oxide nanoparticles can play crucial roles in their properties. Lead oxide may appear in a various forms including nanoplates, nanostars and nanodendrites [17], nanorods [18], nanopowders [19], and nanosheets and nanotubes [20]. The color of tetragonal crystalline structure is red with α-PbO form known as litharge and a bandgap of 1.9–2.2 eV, which has shown to be stable at room temperature, while orthorhombic crystalline structure is yellow with β-PbO form known as massicot and a bandgap of 2.7 eV, which seems to be stable at high temperatures, above 488°C [21]. The conversion phase of α-PbO to β-PbO occurs at approximately 490°C. PbO yellow nanostructure possesses photoconductive properties. α-PbO and β-PbO adopt similar open-packed rock salt-like structures which arise from the stereochemically active lone pair of Pb^{2+} but with different orientations and distortions of PbO_4 pyramid units (**Figure 2**).

Pb atoms are shown as blue spheres and O atoms in red. The distances of adjacent Pb layers are drawn with lines for contrast [22].

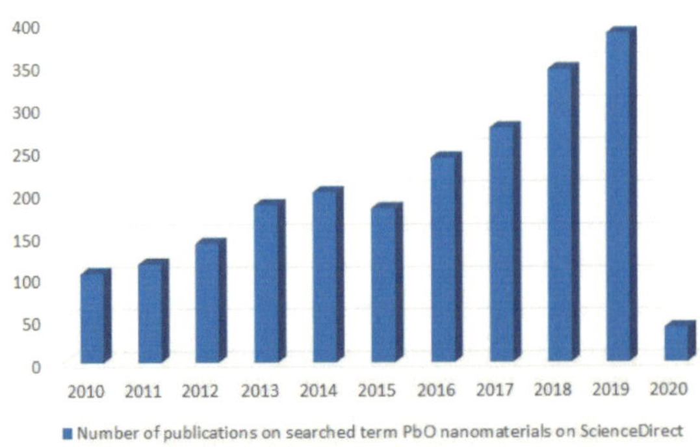

Figure 1.
The results of published papers on ScienceDirect by searching the keyword "PbO nanoparticles."

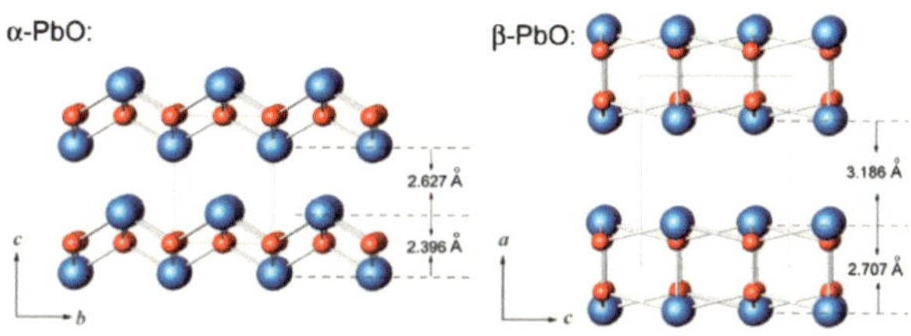

Figure 2.
Crystal structures of α-PbO and β-PbO.

2. Synthesis

In recent years, the synthesis of nanomaterials is an important topic of research in modern material science. For several years, scientists have constantly explored different synthetic methods to synthesize nanoparticles. Lead nanoparticles may be synthetized by three different methods, namely, physical, chemical, and green (biological) methods. Recently, several new routes of synthesis of PbO nanostructures have been used, such as chemical synthesis [23], calcination [24], sol-gel pyrolysis [25, 26], anodic oxidation [27], solvothermal method [16, 28, 29], thermal decomposition [30–32], chemical deposition [33], microwave irradiation technique [34], and green method [35].

Solvothermal synthesis is generally directed to crystal synthesis or crystal growth under high-temperature and high-pressure solvent conditions from substances which are insoluble in normal customary temperature and pressure in an autoclave.

Size-controlled preparation of semiconducting nanostructures is a major challenge in nanoscience and nanotechnology because their essential physiochemical properties and applications are forcefully related to the size-dependent quantum size effect. There has been a great attention of interest in developing various techniques for controlling the nanostructure sizes and shapes. Several studies have shown that in the course of the synthesis, a variety of factors including temperature, concentrations, the interplay between the metal ion precursors and the reducing agent, and adsorption kinetics are strongly influenced on the sizes, shapes, stability, and physicochemical properties of the nanostructures [36, 37].

Mathew and Krishnamurthy [38] in their study in 2018 have shown that the average size of the nanoparticles during chemical synthesis is higher than that during biological synthesis. The size of nanoparticles was found to be 1000 nm, whereas during biological synthesis the particle size was found to be 100–200 nm and above through dynamic light scattering. Using Debye-Scherrer equation, the particle size obtained through chemical synthesis was found to be 180 nm, whereas it was around 78 nm during biological synthesis.

The shape variation of the nanoparticles is important for several applications: oral delivery of therapeutic drugs and diagnostic materials, creation of the antibacterial coatings, and nanotoxicological research. Actually, the size, shape, and surface chemistry of nanoparticles can greatly impact cellular uptake and efficacy of the treatment. The results of the research carried out by Banerjee et al. [39] have shown that spherical nanoparticles in the co-culture have lower cellular uptake compared to rod-shaped nanoparticles regardless of the presence of active targeting moieties. Transport of spherical nanoparticles across the intestinal co-culture was also significantly lower than that of nanorods. This result indicates that nanoparticle-mediated oral drug delivery can be improved using rod-shaped nanoparticles instead of spherical shape.

The laser ablation in water gives the unique ability to produce the colloids (stable water suspensions) of spherical nanoparticles of pure metals and metal oxides. Shur et al. demonstrated the formation of the nonspherical particles at hot water treatment at a temperature above 70°C during laser ablation [40].

Method of synthesis	Source and conditions	Temperature/ time (°C)	Particle size (nm)	Reference
Chemical reactions	$Pb(C_2H_3O_3)_{2,3} \times H_2O$ + 19 M NaOH	90°C	TEM/SEM = 99 XRD = 60	[23]
Calcination	Lead citrate $Pb(C_6H_6O_7)xH_2O$ Under N_2 and air	370°C, 20 min	N_2: 50–60 Air: 100–200	[24]
Thermal decomposition	Lead stearate in octanol	80°C, 2–12 h	60–100 nm	[41]
	Lead hydroxycarbonate Urea and lead nitrate Under microwave	In vacuum at 90°C, 2 h Calcined at 400–600°C, 2 h	α-PbO = 30 β-PbO = 38	[34]
	Lead oxalate (PbC_2O_4)	Precursor: 60°C, 7 h Calcined at 425°C, 3 h	20–30	[42]
Laser ablation	Lead plate that has 99.99% purity	Nd:YAG laser with laser energy of 400 mJ/pulse, with 1064 nm wavelength, 9 ns		[43]
	Pb of 99.99% purity		2D NPs and 3D NPs Micron and submicron diameter	[40]
Sol-gel	Lead nitrate Gelatin	60°C, 30 min 90°C Calcined at: 500–600°C under air, 2 h	TEM = 45	[44]
	Lead acetate (2.5 wt%) Polyvinyl alcohol (8 wt%) H_2O/EtOH (40/60V/V)	80°C Calcined at 500°C, 5 h	18–43	[26]
Green/biosynthesis	$Pb(NO_3)_2$ Gelatin	90°C, 2 h Calcined at 450–600°C, 90 min	10–20	[45]
	Lead acetate, *Papaver somniferum*	60°C, 2 h Calcined at 500°C, 2 h	23 ± 11	[46]
	Lead nitrate Extract *cyminum*	Microwave 700 W 5 min	40–150 87 XRD = 35.8–45.41	[35]

Table 1.
An overview of the methods of synthesis and the main properties of Pb-NPs.

Table 1 shows an overview of the methods of synthesis and the main properties of Pb-NPs.

It is noticeable from **Table 1** that the particle size decreases starting from chemical synthesis to sol-gel and green synthesis. It is possible to note that in the nitrogen atmosphere, the particle size is smaller (50–60 nm) than that done in air (100–200 nm). By changing the calcination temperature, it is possible to tune the ratio α-PbO and β-PbO and consequently change the final properties of synthesized material. By using microwave instead of heating, it is possible significantly to reduce the reaction time and obtain the final material with different properties.

2.1 Chemical synthesis of lead oxide nanoparticles

Alagar et al. [23] have synthesized lead oxide (PbO) semiconductor nanoparticles by chemical synthesis. During this synthesis, they prepared 1 M aqueous solution of $Pb(C_2H_3O_2)_{2.3}$ H_2O in deionized water and heated up to 90°C. They mixed this solution with an aqueous solution of 19 M NaOH and stirred vigorously. Immediately after mixing the two solutions, the cloudy solution is observed and then turned peach and finally deep orange red. At this moment, the reaction was stopped, and the formation of precipitate was evident. The supernatant was decanted, filtered on a Buchner funnel, washed with deionized water repeatedly, and dried overnight in a drying oven at 90°C. The sample was removed and lightly crushed in a mortar and pestle.

2.2 Calcination

Li et al. [24] have prepared nanostructural lead oxide by decomposition of lead citrate in inert (nitrogen) and in the air. The lead citrate $[Pb(C_6H_6O_7)\cdot H_2O]$, a precursor, was synthesized through leaching of spent lead-acid battery paste in citric acid system. In this citric acid leaching system, $C_6H_8O_7\cdot H_2O$ was used to supply the citric acid aqueous solution with distilled water for the balance of pH and for leaching and crystallization. The citrate particle was in the shape of sheets with the length of 10–20 μm, width of 2–10 μm, and thickness of 0.5 μm.

The major decomposition products were orthorhombic phase β-PbO, metallic Pb, and elemental C, when the calcination was carried out in nitrogen gas. The ratio of β-PbO to Pb increased with the increasing temperature. The size range of the particle was 50–60 nm with a spherical shape. Lead citrate could be calcined completely within 20 min at 370°C in air conditions. The particle size obtained was in range of 100–200 nm. It was found to compose mainly of β-PbO with a small part of α-PbO and Pb.

2.3 Synthesis by sol-gel method

In 2014 Yousefi and co-workers [44] have synthetized in a gelatin medium single-crystal, lead oxide nanoparticles (PbO-NPs) via the sol-gel method. In order to terminate the growth of the PbO-NPs and stabilize them, the long-chain gelatin compounds were utilized. The calcination of the gel obtained was done at 500, 550, and 600°C for 1 h. Miri et al. [45] have synthetized PbO-NPs through the utilization of gelatin as a green stabilizer.

2.4 Synthesis by thermal decomposition

The thermal decomposition of lead stearate in octanol was used by Akimov and co-workers [41] for the synthesis of lead nanoparticles. They showed that by modifying the concentration of lead stearate in octanol and changing the thermolysis, time is possible to control the lead particle size. An organic coat composed of decomposition products of lead stearate prevents the particles from oxidation in air and favors their dissolution in organic solvents. Li et al. [34] have prepared PbO-NPs by thermal decomposition of lead hydroxycarbonate synthesized under microwave irradiation. Urea and lead nitrate were used as starting materials.

2.5 Green synthesis

Fundamentally, the main advantage of green method for the synthesis of nanoparticles is easy, efficient, and eco-friendly [38] in comparison to chemical-mediated or microwave-mediated green synthesis. The chemical synthesis involves toxic solvents, high pressure and energy, and high-temperature conversion, whereas microbe-involved synthesis is not feasible industrially due to its lab maintenance. Green biologically based methods which are safe, inexpensive, and an environment-friendly alternative usually imply the use of microorganisms and plants to synthesize nanoparticles [38, 47].

Nanoparticles can be synthesized from a wide variety of biological entities such as actinomycetes, algae, bacteria, fungus, plants, viruses, and yeast [48].

Gandhi et al. [35] have synthesized lead nanoparticles using the extract of *Cuminum cyminum* seed powder extract. The synthesized nanoparticles showed efficient antimicrobial activities against bacteria and pathogenic fungi. Similarly, the synthesized nanoparticles showed efficient anti-algal activity against spirulina culture.

3. Methods of characterization

For the structural characterization of lead oxide nanoparticles, several techniques are used including UV–visible spectrophotometry, FTIR spectroscopy, X-ray diffraction (XRD), energy-dispersive X-ray spectroscopy (EDX), scanning electron microscopy (SEM), thermogravimetric analysis (TGA), and thermogravimetric-differential thermal analysis (TG-DTA).

3.1 UV–visible studies

The optical properties of the PbO-NPs usually are studied by using a UV-visible (UV-vis) spectrophotometer. Many scientific papers indicate the presence of the absorption peaks in the visible region on the UV-vis absorption spectra of the PbO-NPs, which are attributed to the bandgap of the PbO-NPs with an orthorhombic phase [44]. The energy bandgap value of the prepared sample is 3.82 eV estimated using Tauc relation [23].

In the study of Miri et al. [45], the electronic spectra of the synthesized PbO-NPs have displayed a wide peak in 250–260 nm at the calcination temperatures of 450, 500, 550, and 600°C. They have discovered that by increasing the temperature from 450 to 600°C, the wavelength is repositioned from 252 to 260 nm, and the alpha form of PbO converts to beta.

The particle size of synthesized nanoparticles may be determined through the bandgap of electron spectrum by using the Tauc equation:

$$(\alpha h\upsilon) = A\,(h\upsilon - E_g)^n \quad (1)$$

where A is the absorption coefficient, hυ is the photon energy, and E_g is the bandgap energy.

They have calculated values of bandgap energy for the synthesized PbO-NPs which were 4.38, 4.39, 4.30, and 4.02 eV at the calcination temperatures of 450, 500, 550, and 600°C, respectively. It is revealed that increasing the bandgap energy results in a smaller particle size. The calculated values of bandgap energy are much larger than the ordinary lead oxide bandgap indicating that the synthesized powder is in nanoscale and its alpha form is smaller than the beta. The bandgap energy of α-PbO and β-PbO is 1.92 and 2.7 eV.

3.2 Fourier transform infrared (FTIR)

The chemical bonding of lead oxide NPs may be investigated by FTIR. According to Alagar et al. [23], two very sharp peaks at 466.74 and 557.39 cm^{-1} on the FTIR spectra indicate the presence of lead and oxide.

FTIR spectrum for lead oxide nanoparticles synthesized by laser ablation of lead target immersed in deionized water by using pulsed Nd:YAG laser with laser energy of 400 mJ/pulse and different laser pulses has shown different transmission peaks [43]. The appearance of the transmission peak at 460cm^{-1} indicates the presence of (PbO) stretching vibration mode; the peaks located at (763.84, 1057.03) cm^{-1} and the peaks at (3877.05, 3742.03) cm^{-1} are corresponding to the bonding of (O—H) and harmonics of H—OH stretching bonding modes of water, while the infrared peaks located at (2856.67, 2918.40, 1743.1) and 1658.84 cm^{-1} are related to the (C=O) stretching vibration modes that refer to little contribution of CO_2 dissolution from air contain.

The broad peak around 3400 cm^{-1} corresponds to (O—H) stretching vibrations originating from small amount of ethanol used for washing the samples. A sharp peak at 1705 cm^{-1} is assigned to vibrations of the carbonyl group (C=O), while another peak at 1395 cm^{-1} is recognized as stretching vibration of carboxyl group (C—O). Metal-oxygen (M—O) stretching vibrations were identified by the peak presence in the region 300–900 cm^{-1}.

3.3 X-ray diffraction (XRD)

X-ray powder diffraction (XRD) is a rapid analytical technique primarily used for phase identification of a crystalline material and can provide information on unit cell dimensions. The average grain size of lead oxide nanoparticles synthetized by chemical synthesis was determined using *Debye–Scherrer equation*.

The *Debye-Scherrer* equation is [49]:

$$D_{hkl} = K\lambda / (B_{hkl} \cos\theta) \tag{2}$$

where,

D_{hkl} is the crystallite size in the direction perpendicular to the lattice planes.

$_{hkl}$ are the Miller indices of the planes.

K is a numerical factor frequently referred to as the crystallite-shape factor.

λ is the wavelength of the X-rays.

B_{hkl} is the width (full-width at half-maximum) of the X-ray diffraction peak in radians.

θ is the Bragg angle.

Specific surface area is 10.52 m^2/g, surface area of particle is 11,304 nm^2, particle volume is 113,040 nm^3, and density value is 9.5 g/cm^3 [23]. Since the crystallinity index is ≈1.65, the sample is polycrystalline. The particle size (obtained from either TEM or SEM) is 99 nm, whereas that calculated from XRD is 60 nm [23].

In the research carried out by Yousefi et al. [44], XRD patterns indicated a phase transformation from tetragonal to orthorhombic as the calcination temperature increased from 500 to 600°C for 1 h. These Pb-NPs were synthesized by sol-gel method, and the SEM and TEM images showed that the nanoparticles started to form at a temperature range of 550–600°C. In addition, the TEM results showed that the average particle size of the nanoparticles was ~45 nm. It is possible to note that by changing the method of synthesis, i.e., from chemical to sol-gel method, the nanoparticle size decreases from 60 [23] to 45 nm [44] and even to around 10–20 nm [45].

In the research carried out by Nafees et al. [42], X-ray diffraction (XRD) patterns were recorded in the 2θ range of 20–75 using a step size of $0.05°s^{-1}$.

XRD pattern of PbO is indexed as tetragonal; a very small peak present at 30.25 (2θ) shows the presence of orthorhombic phase. Long heating time required to decompose PbC_2O_4 precursor results in the phase change and a corresponding appearance of the additional peak. Average crystallite sizes for PbO and PbS were found to be around 20–30 nm as calculated by *Debye-Scherrer equation.*

3.4 Energy-dispersive X-ray spectroscopy (EDX)

For the identification and quantification of elemental compositions in a very small amount of material (few cubic micrometers of the sample), energy-dispersive X-ray spectroscopy (EDX) is used. The method is based on the excitation of the atoms on the surface by the electron beam, emitting specific wavelengths of X-rays that are the characteristic of the atomic structure of the elements. An energy-dispersive detector (a solid-state device that discriminates among X-ray energies) can analyze these X-ray emissions. Appropriate elements are assigned, yielding the composition of the atoms on the specimen surface [50].

3.5 Scanning electron microscopy (SEM)

SEM technique based on electron scanning provides all available information about the nanoparticles at nanoscale level. This technique is useful not only to study the morphology of the nanomaterials but also the dispersion of nanoparticles in the bulk or matrix. In the research carried out by Nafees and co-workers [42], lead oxalates (PbC_2O_4) were used as a precursor for the synthesis of nanocrystalline lead oxide (PbO) by thermal decomposition.

3.6 Thermogravimetric analysis (TGA) and thermogravimetric-differential thermal analysis (TG-DTA)

Generally, the thermal analysis of TGA-DTA of the sample serves to obtain the proper calcination temperature to reach the pure phase. A better understanding of oxidation temperature/resistance and thermal stability of PbO and PbS is crucial for realizing the potential applications these materials promise. Still, sufficient literature on these issues is not available. Thermal stability and various transitions/reactions occur in nanostructured PbO and PbS during heat treatment in air and inert atmosphere [42]. In the research carried out by Miri et al. [45], the thermal analysis of TGA-DTA of the synthesized sample was performed in the temperature interval of 20–1000°C in an atmospheric air. On the DTA curve, the dehydration and discharge of crystalline and absorbed water were attributed in the temperature range of 80–235°C. In the same temperature range on the TGA curve, the first weight loss of 35% was observed. In the temperature interval of 235–460°C, the second weight loss of 16% was attributed to the decomposition of gelatin chemical bond. In the temperature interval of 460–480°C, the third weight loss of 7% was related to lead oxidation, and the last weight loss of 6% was associated with the final oxidation of lead. Their study has discovered that PbO-NPs were stable in a temperature range between 550 and 880°C, since no weight changes were observed. In the same study, for the PbO, DSC/TGA curves were recorded from room temperature to 1175°C in air with the ramp rate of 10°C min^{-1}. They observed that a mass loss of almost 2% occurs up to 500°C. As observed from differential thermogravimetric (DTGA) curve, dehydration takes place first resulting in the removal of water of crystallization around 100°C, while a hump between 200 and 250°C corresponds to the endothermic nature of transition confirmed by DSC curve. The thermal decomposition from PbO_2 to PbO occurs in a number of steps from 290 to 650°C accompanied by endothermic mass losses in designated temperatures. In DSC curve at 880°C, a sharp endothermic peak was observed indicating on the melting of PbO without the involvement of any mass loss as supported by TGA and DTGA curves. Probably because of the partial evaporation of very small PbO nanoparticles, another endothermic mass loss was observed around 1100°C.

Excellent thermal stability of PbO nanoparticles in a range from room temperature to 1175°C was attributed since only 2.7% of mass loss was recorded.

4. Applications

Lead oxide is a semiconductor nanostructure that has important applications in storage batteries, pigments, ceramics, and glass industry.

4.1 Storage batteries

Lead oxide has wide industrial applications as a basic material of electrode (anode and cathode) active mass in lead-acid batteries.

The lead-acid battery market is divided into industrial, data centers, telecom, oil and gas, and others. The industrial segment includes construction, metal and mining, chemical and pharmaceutical, and food and beverage industries. Due to the growth of the IT sector, Asia Pacific is focusing on increasing the number of data centers installed across the countries. It is expected that the lead-acid batteries will be used as a backup power solution in the data centers owing to their functionality across a wide temperature range. According to projections, the global lead-acid battery market could reach USD 52.5 billion by 2024 from an estimated USD 41.6 billion in 2019, at a CAGR of 4.7% during the forecast period. The booming telecommunication sector and expanding data industry require a cost-effective battery storage solution to provide for backup power [51].

Low-cost manufacture, simplicity of design, reliability, and relative safety are some of the advantages of lead-acid battery comparing to other electrochemical systems. Relatively, good specific power has caused a widespread use of lead-acid batteries in starting, lightening, and ignition of engine purposes for vehicular applications. Because of the advantages of lead-acid batteries, there is a big interest to improve and develop lead oxide nanostructures to obtain more discharge capacity and more life cycle [52].

In formation process, lead oxide can be converted to spongy lead in anode and lead dioxide in cathode. Some advantages of lead-acid batteries include low-cost manufacture, simplicity of design, reliability, and relative safety when compared to other electrochemical systems [26]. Lead oxide nanoparticles (PbO-NPs) have well-known applications in lead storage batteries. Every vehicle which has a lead-acid battery uses 5–10 kg of lead oxide in its production. This represents vast uses of lead oxide on a worldwide basis. Batteries designed for multiple charge–discharge cycles are *traction batteries*, and red lead (Pb_3O_4) oxide is used for their production. Traction batteries are used to power all manners of electrical vehicles including fork-lift trucks, golf carts, milk floats, etc. For standby power applications in case of main supply failure in hospitals, computer installations, and telecom networks, *stationary batteries* are required. Again, red lead (Pb_3O_4) oxide is common in these units.

4.2 Sensors

Li et al. [34] have prepared PbO-NPs ($Pb_3(CO_3)_2(OH)_2$, α-PbO, and β-PbO) by using microwave irradiation technique. The reaction time was only 17 min and was significantly shortened from 4 h of conventional water bath heating method. Hence, the use of microwave irradiation technique is a fast, convenient, mild, energy-efficient, and environment-friendly route to produce the lead nanoparticles. For the preparation of the Pb(II)-selective electrode, cellulose acetate-acetone method which is simple, economical, and highly efficient to form membrane was used. α-PbO nanoparticle obtained by calcining $Pb_3(CO_3)_2(OH)_2$ at 400°C also was used. Experiment results indicate that this electrode exhibited a Nernstian response for Pb^{2+} ion in a linear range of 2.5×10^{-5} mol L^{-1} to 1.0×10^{-1} mol L^{-1}. High selectivity for some metal ions, especially for the interference ions, such as Cu^{2+}, Ag^{+}, and Hg^{2+}, were obtained.

4.3 Pigments

The major lead pigment is the red lead (Pb_3O_4), which is used principally in ferrous metal protective paints, and litharge, a bright yellow form of lead monoxide, still finds applications in the production of yellow pigments. The colors that contain lead-based pigments seem to have interesting properties including rustproof, antibacterial, and anti-algae, which are extensively employed in shipbuilding, construction skeleton, and road construction [45].

4.4 Glass

The clarity and density of leaded glass crystal can be enhanced by the presence of 24–28% of lead oxide. The presence of lead oxide in television tube may reduce harmful radiation. Lead oxide in the optical fibers enhances the refractive index.

Lead oxides are also used in special optical glasses, X-ray protection glasses, etc. A lead oxide in glazes and enamels enhances the thermal, color, and wear properties of such coatings.

4.5 Stabilizers

Lead oxide as lead stabilizers in polyvinylchloride (PVC) processing is used. They improve the thermal stability of PVC, allowing high-temperature processing, and the electrical and UV resistance properties.

4.6 Ceramics

Glassy materials that experience controlled crystallization, usually achieved by a heating process in the presence of nucleating agents, are named glass-ceramics.

Almeida et al. [53] have shown that lead oxide is an important glass modifier not only for affecting the chemical and mechanical stabilities of glasses but also for improving their thermal and optical properties. They have designed and controlled β-PbO and $3PbO{\cdot}H_2O$ crystalline phases in a lead borate glass using femtosecond-direct laser writing (fs-DLW) followed by chemical etching at room temperature. They demonstrated that for the glass crystallization, the etching in aqueous KOH solution was responsible, whereas the grooves produced by fs-laser pulses enable the selective crystallization in a predetermined 2D pattern. By using this method, phase transformation at the micrometer scale can be controlled.

4.7 Anticancer and antimicrobial drugs

Several researches indicate that lead oxide nanoparticles may act as significant anticancer and antimicrobial drugs. Miri et al. [45] have synthesized lead oxide nanoparticles through the utilization of gelatin as a stabilizer and studied its cytotoxic effects on Neueo2A cancer cell line. The results of this research indicated that concentrations of PbO-NP under 30 μg/mL have insignificant toxicity. Muhammad et al. [46] have synthesized lead oxide (PbO) and iron oxide (Fe_2O_3) nanoparticles starting from *Papaver somniferum L. Papaver somniferum*, also known as opium poppy and as an important member of *Papaveraceae* family. Poppy plant is well recognized for its diverse pharmacological potentials. Poppy seeds have the analgesic potency and are used as source of oil. Both NPs also showed considerable total antioxidant potential, free radical scavenging potential, and reducing power. PbO showed more potent anticancer activity than Fe_2O_3 NPs. The antibacterial property results show that the antibacterial activity of the lead oxide NPs was inversely proportional to the size of the nanoparticles in both Gram-negative and Gram-positive. It has been found that Gram-positive bacteria possess greater sensitivity to the lead oxide nanoparticles compared with Gram-negative bacteria [43].

5. Toxicity

Lead oxide nanoparticles are graded as toxic and dangerous for the humans and environment. Lead and its compounds are extremely poisonous. Lead is a heavy metal whose toxicity is based on its selective and practically irreversible retention in the nerve cells. In them, it affects the energy balance, the signal transmission, and the flow of sodium, potassium, and calcium ions. A lot of lead is used as tetraethyl lead compound as a petro-fuel additive. With the combustion of gasoline, the lead enters the atmosphere and reaches the respiratory chain or the drinkable water in a form of very fine particles. Some of the lead is absorbed by plants and animals, so it reintroduces food into the human body. The major target for lead toxicity is the nervous system. Poisoning by the lead happens after the consumption of food or water polluted with lead but also after unplanned ingestion of contaminated soil, dust, or lead-based paint [54].

Nanoscale lead dioxide particles ($nPbO_2$) formed inside lead-bearing pipes or lead-containing faucets in drinking water distribution systems can release toxic lead ions, causing drinking water contamination. Ng et al. [55] used adult medaka fish (*Oryzias latipes*) as an in vivo model to investigate the bioavailability and toxicity of $nPbO_2$ in an intact animal. Both types of PbO_2 particles, nanoscale [$nPbO_2$] and microscale bulk [$bPbO_2$], were chemically stable in dechlorinated tap water with low water solubility. However, both $nPbO_2$ and $bPbO_2$ could be reductively dissolved into $Pb(II)_{aq}$ in both the intestine (major uptake route) and gills of the fish, thereby enhancing hepatic Pb accumulation. The Pb content was greater in the gills, liver, and brain with nanoscale than in the microscale bulk PbO_2 particles. This in vivo evidence implies the possibility of increased risk of exposure to Pb dissolution from PbO_2 particles in the digestive system via drinking water, which can enhance the bioavailability of Pb uptake and toxicity in humans.

Recent studies have indicated that even lower blood Pb toxicity causes several metabolic, neurological, and behavioral disorders, as well as may be associated with impairment in psychological progress, diminished skeletal growth, and disturbances in cardiovascular function [56]. Lead poisoning is one of the most common professional diseases in industry. De facto, in industrial plants, lead is introduced into the body by inhalation of polluted air, in which lead or its compounds are contained in the form of smoke, dust, or vapor. The quantity of lead absorbed and the degree of poisoning depend on the concentration of lead in the air, its physical or chemical state with respect to the suitability for absorption, and the time of exposure of a worker in the contaminated space. In many industrially advanced countries, toxicity limits are set for the workers handling various toxic and hazardous materials widely used in the manufacturing batteries. In general, a worker's maximum allowable lead inhalation during an 8-h working period is 0.15 mg/m^3. The toxic effect of lead is based on its selective and practically irreversible retention in the nerve cells. Lead affects the energy balance, the signal transmission, and the flow of sodium, potassium, and calcium ions.

6. Conclusions

Nanotechnology is an emerging field in the current global market, and the use of nanoparticles has shown enormous applications in every industrial sector due to its unique properties. The number of published scientific papers clearly shows that interest in PbO-NPs has been increasing year by year over the last 10 years. The lead oxide nanoparticles are one of the most industrially used metal nanoparticles.

The global lead-acid battery market is projected to reach USD 52.5 billion by 2024 because the booming telecommunication sector and expanding data industry require a cost-effective battery storage solution to provide for backup power.

Different ways of synthesis of PbO nanoparticles were present through the chapter with emphasis on different final properties of synthesized material as well as disadvantages of chemical and physical route, including generation of toxic chemicals which cause damage to the environment as well as humans and advantages of green synthesis. Therefore, to overcome the disadvantages of chemical and physical methods, the material scientists have proposed the new "green" synthesis pathways of nanoparticles which are a simple, fast, economical, convenient, energy-efficient, and environment-friendly way to produce lead oxide nanoparticles with multifarious applications. This is possible by changing the method of synthesis (process parameters, precursor concentration, temperature, atmosphere, the way of heating) to tune the final properties of synthesized lead oxide nanoparticles (particle size, optical, mechanical, catalytic properties).

The advantage of laser ablation is underlined because in water it is possible to produce the colloids of spherical nanoparticles of pure metals and metal oxides which are important for the creation of the antibacterial coatings and nanotoxicological research. However, lead oxide nanoparticles are graded as toxic and dangerous for the human and environment, and therefore there is an urgent need to develop new approaches and standardized test procedures to study the potential hazardous effect of nanoparticles on human health and environment.

Conflict of interest

The author declares that she has no conflict of interest.

Author details

Amra Bratovcic
Faculty of Technology, Department of Physical Chemistry and Electrochemistry, University of Tuzla, Tuzla, Bosnia and Herzegovina

*Address all correspondence to: amra.bratovcic@untz.ba

References

[1] Bratovcic A. Different applications of nanomaterials and their impact on the environment. SSRG International Journal of Material Science and Engineering. 2019;**5**:1-7

[2] Feynman RP. There's plenty of room at the bottom. Engineering and Science. 1960;**22**:22-36

[3] Laurent S, Forge D, Port M, Roch A, Robic C, Vander Elst L, et al. Magnetic iron oxide nanoparticles: Synthesis, stabilization, vectorization, physicochemical characterizations, and biological applications. Chemical Reviews. 2010;**110**:2574-2574. DOI: 10.1021/cr900197g

[4] Singh P, Kumari K, Vishvakrma VK, Mehrotra GK, Chandra R, Kumar D, et al. Metal NPs (Au, Ag, and Cu): Synthesis, stabilization, and their role in green chemistry and drug delivery. In: Singh R, Kumar S, editors. Green Technologies and Environmental Sustainability. Cham: Springer; 2017. pp. 309-337. DOI: 10.1007/978-3-319-50,654-8_14

[5] De Gaetano F, Ambrosio L, Raucci MG, Marotta A, Catauro M. Sol-gel processing of drug delivery materials and release kinetics. Journal of Materials Science: Materials in Medicine. 2005;**16**(3):261-265. DOI: 10.1007/s10856-005-6688-x

[6] Crabtree JH, Burchette RJ, Siddiqi RA, Huen IT, Hadnott LL, Fishman A. The efficacy of silver-ion implanted catheters in reducing peritoneal dialysis-related infections. Peritoneal Dialysis International: Journal of the International Society for Peritoneal Dialysis. 2003;**23**(4):368-374

[7] Królikowska A, Kudelski A, Michota A, Bukowska J. SERS studies on the structure of thioglycolic acid monolayers on silver and gold. Surface Science. 2003;**532-535**:227-232. DOI: 10.1016/S0039-6028(03)00094-3

[8] Zhao G, Stevens SEJ. Multiple parameters for the comprehensive evaluation of the susceptibility of *Escherichia coli* to the silver ion. Biometals. 1998;**11**(1):27-32. DOI: 10.1023/a:1009253223055

[9] Theivasanthi T, Alagar M. X-ray diffraction studies of copper Nanopowder. Archives of Physics Research. 2010;**1**(2):112-117

[10] Singh M, Manikandan S, Kumaraguru AK. Nanoparticles: A new technology with wide applications. Research Journal of Nanoscience and Nanotechnology. 2011;**1**(1):1-11. DOI: 10.3923/rjnn.2011.1.11

[11] Sonmez M, Kumar R. Leaching of waste battery paste components. Part 1: Lead citrate synthesis from PbO and PbO_2. Hydrometallurgy. 2009;**95**:53-60. DOI: 10.1016/j.hydromet.2008.04.012

[12] Sljukić B, Banks CE, Crossley A, Compton RG. Lead(IV) oxide-graphite composite electrodes: Application to sensing of ammonia, nitrite and phenols. Analytica Chimica Acta. 2007;**587**:240-246

[13] Senvaitiene J, Smirnova J, Beganskiene A, Kareiva A. XRD and FTIR characterisation of lead oxide-based pigments and glazes. Acta Chimica Slovenica. 2007;**54**:185-193

[14] Jaffe B, Roth R, Marzullo S. Properties of piezoelectric ceramics in the solid-solution series lead titanate-lead zirconate-lead oxide: Tin oxide and lead titanate-lead hafnate. Journal of Research of the National Bureau of Standards. 1955;**55**(5):239-254

[15] Dumbaugh WH, Lapp JC. Heavy-metal oxide glasses. Journal of

the American Ceramic Society. 1992;**75**(9):2315-2326. DOI: 10.1111/j.1151-2916.1992.tb05581.x

[16] Tayebee R, Maleki B, Sabeti M. A new simple method for the preparation of PbO nanoparticles and implementation of an efficient and reusable catalytic system for the synthesis of 1H-pyrazolo[1,2-b]phthalazine-5,10-diones. Journal of the Iranian Chemical Society. 2017;**14**(6):1179-1188. DOI: 10.1007/s13738-017-1068-2

[17] Chen KC, Wang CW, Lee YI, Liu HG. Nanoflakes and nanostars of β-PbO formed at the air/water interface. Colloids and Surfaces A: Physicochemical and Engineering Aspects. 2011;**373**(1-3):124-129. DOI: 10.1016/j.colsurfa.2010.10.035

[18] Ghasemi S, Mousavi MF, Shamsipur M, Karami H. Sonochemical-assisted synthesis of nano-structured lead dioxide. Ultrasonics Sonochemistry. 2008;**15**(4):448-455. DOI: 10.1016/j.ultsonch.2007.05.006

[19] Kashani-Motlagh MM, Karami Mahmoudabad M. Synthesis and characterization of lead oxide nano-powders by sol-gel method. Journal of Sol-Gel Science and Technology. 2011;**59**(1):106-110. DOI: 10.1007/s10971-011-2467-y

[20] Shi L, Xu Y, Li Q. Controlled growth of lead oxide nanosheets, scrolled nanotubes, and nanorods. Crystal Growth Design. 2008;**8**(10):3521-3525. DOI: 10.1021/cg700909v

[21] Schottmiller JC. Photoconductivity in tetragonal and orthorhombic lead monoxide layers. Journal of Applied Physics. 1966;**37**(9):3505-3510. DOI: 10.1063/1.1708890

[22] Wang Y, Lin X, Zhang H, Wen T, Huang F, Li G, et al. Selected-control hydrothermal growths of α- and β-PbO crystals and orientated pressure-induced phase transition. CrystEngComm. 2013;**15**:3513-3516. DOI: 10.1039/c2ce26162f

[23] Alagar M, Theivasanthi T, Kubera Raja A. Chemical synthesis of nano-sized particles of lead oxide and their characterization studies. Journal of Applied Sciences. 2012;**12**(4):398-401. DOI: 10.3923/jas.2012.398.401

[24] Li L, Zhu X, Yang D, Gao L, Liu J, Vasant Kumar R, et al. Preparation and characterization of nano-structured lead oxide from spent lead acid battery paste. Journal of Hazardous Materials. 2012;**203-204**:274-282

[25] Mahmoudabad MK, Kashani-Motlagh MM. Synthesis and characterization of PbO nanostructure and NiO doped with PbO through combustion of citrate/nitrate gel. International Journal of the Physical Sciences. 2011;**6**(24):5720-5725

[26] Karami H, Ghamooshi-Ramandi M. Synthesis of sub-micro and nanometer sized Lead oxide by sol-gel pyrolysis method and its application as cathode and anode of Lead-acid batteries. International Journal of Electrochemical Science. 2013;**8**:7553-7564

[27] Singh DP, Srivastava ON. Synthesis of micron-sized hexagonal and flowerlike nanostructures of lead oxide (PbO_2) by anodic oxidation of lead. Nano–Micro Letters. 2011;**3**(4):223-227. DOI: 10.1007/BF03353676

[28] Gao P, Liu Y, Bu X, Hu M, Dai Y, Gao X, et al. Solvothermal synthesis of α-PbO from lead dioxide and its electrochemical performance as a positive electrode material. Journal of Power Sources. 2013;**242**:299-304. DOI: 10.1016/j.jpowsour.2013.05.077

[29] Jia B, Gao L. Synthesis and characterization of single crystalline PbO nanorods via a facile hydrothermal

method. Materials Chemistry and Physics. 2006;**100**(2-3):351-354. DOI: 10.1016/j.matchemphys.2006.01.012

[30] Salavati-Niasari M, Mohandes F, Davar F. Preparation of PbO nano-crystals via decomposition of lead oxalate. Polyhedron. 2009;**28**(11):2263-2267. DOI: 10.1016/j.poly.2009.04.009

[31] Behnoudnia F, Dehghani H. Synthesis and characterization of novel three-dimensional-cauliflower-like nanostructure of lead (II) oxalate and its thermal decomposition for preparation of PbO. Inorganic Chemistry Communications. 2012;**24**:32-39. DOI: 10.1016/j.inoche.2012.07.031

[32] Cimen E, Gumus I, Arslan H. The role of intermolecular interactions in the assembly of zinc(II) and Lead(II) complexes containing carboxylate ligand and their conversion to metal oxides. Journal of Molecular Structure. 2018;**1166**:397-406. DOI: 10.1016/j.molstruc.2018.04.024

[33] Mythili N, Arulmozhi K. Characterization studies on the chemically synthesized α and βPhase PbO nanoparticles. International Journal of Scientific and Engineering Research. 2014;**5**(1):412-416

[34] Li S, Yang W, Chen M, Gao J, Kang J, Qi Y. Preparation of PbO nanoparticles by microwave irradiation and their application to Pb(II)-selective electrode based on cellulose acetate. Materials Chemistry and Physics. 2005;**90**:262-269. DOI: 10.1016/j.matchemphys.2004.02.022

[35] Gandhi N, Sirisha D, Asthana S. Microwave mediated green synthesis of lead (Pb) nanoparticles and its potential applications. International Journal of Engineering Sciences and Research Technology. 2018;**7**(1):623-644. DOI: 10.5281/zenodo.1161701

[36] Alexandridis P. Gold nanoparticle synthesis, morphology control, and stabilization by functional polymers. Chemical and Engineering Technology. 2011;**34**(1):15-28. DOI: 10.1002/ceat.201000335

[37] Wang L, Chen X, Zhan J, Chai Y, Yang C, Xu L, et al. Synthesis of gold nano and microplates in hexagonal liquid crystals. The Journal of Physical Chemistry B. 2005;**109**(8):3189-3194. DOI: 10.1021/jp0449152

[38] Mathew BB, Krishnamurthy NB. Evaluation of lead oxide nanoparticles synthesized by chemical and biological methods. Journal of Nanomedicine Research. 2018;**7**(3):195-198. DOI: 10.15406/jnmr.2018.07.00187

[39] Banerjee A, Qi J, Gogoi R, Wong J, Mitragotri S. Role of nanoparticle size, shape and surface chemistry in oral drug delivery. Journal of Controlled Release. 2016;**238**:176-185. DOI: 10.1016/j.jconrel.2016.07.051

[40] Shur VY, Gunina EV, Esin AA, Shishkina EV, Kuznetsov DK, Linker EA, et al. Influence of hot water treatment during laser ablation in liquid on the shape of PbO nanoparticles. Applied Surface Science. 2019;**483**:835-839. DOI: 10.1016/j.apsusc.2019.03.303

[41] Akimov DV, Andrienko OS, Egorov NB, Zherin II, Usov VF. Synthesis and properties of lead nanoparticles. Russian Chemical Bulletin, International Edition. 2012;**61**(2):225-229. DOI: 10.1007/s11172-012-0032-7

[42] Nafees M, Ikram M, Ali S. Thermal stability of lead sulfide and lead oxide nano-crystalline materials. Applied Nanoscience. 2017;**7**:399-406. DOI: 10.1007/s13204-017-0578-7

[43] Rashed HH, Fadhil FA, Hadi IH. Preparation and characterization of lead oxide nanoparticles by laser ablation as antibacterial agent. Baghdad Science

Journal. 2017;**14**(4):801-807. DOI: 10.21123/bsj.2017.14.4.0801

[44] Yousefi R, Khorsand Zak A, Jamali-Sheini F, Ming Huang N, Jefrey Basirun W, Sookhakian M. Synthesis and characterization of single crystal PbO nanoparticles in a gelatin medium. Ceramics International Part A. 2014;**40**(8):11699-11703. DOI: 10.1016/j.ceramint.2014.03.180

[45] Miri A, Sarani M, Hashemzadeh A, Mardani Z, Darroudi M. Biosynthesis and cytotoxic activity of lead oxide nanoparticles. Green Chemistry Letters and Reviews. 2018;**11**(4):567-572. DOI: 10.1080/17518253.2018.1547926

[46] Muhammad W, Khan MA, Nazir M, Siddiquah A, Mushtaq S, Hashmi SS, et al. *Papaver somniferum* L. mediated novel bioinspired lead oxide (PbO) and iron oxide (Fe_2O_3) nanoparticles: In-vitro biological applications, biocompatibility and their potential towards HepG2 cell line. Materials Science and Engineering: C. 2019;**103**:109740. DOI: 10.1016/j.msec.2019.109740

[47] Shah M, Fawcett D, Sharma S, Tripathy SK, Poinern GE. Green synthesis of metallic nanoparticles via biological entities. Materials. 2015;**8**:7278-7308. DOI: 10.3390/ma8115377

[48] Thakkar KN, Mhatre SS, Parikh RY. Biological synthesis of metallic nanoparticles. Nanomedicine. 2010;**6**(2):257-262. DOI: 10.1016/j.nano.2009.07.002

[49] Holzwarth U, Gibson N. The Scherrer equation versus the "Debye–Scherrer equation". Nature Nanotechnology. 2011;**6**:534. DOI: 10.1038/nnano.2011.145

[50] Ebnesajjad S. Surface and material characterization techniques. In: Surface Treatment of Materials for Adhesive Bonding. 2nd ed. Oxford, GB and Waltham, USA: Elsevier; 2014. pp. 39-75. DOI: 10.1016/B978-0-323-26,435-8.00004-6

[51] Lead Acid Battery Market. Available from: https://www.marketsandmarkets.com/Market-Reports/lead-acid-battery-market-161,171,997.html?gclid=CjwKCAiAx_DwBRAfEiwA3vwZYiEesj31IcKlfbxHx_no4V7GbUiuckyKiINXcVrJ_eW6EUR5kM9RqhoCwlAQAvD_BwE [Accessed: 13 January 2020]

[52] Shah MA. Lead oxide (PbO) nanoparticles prepared by a new technique for biomedical applications. International Journal of Biomedical Nanoscience and Nanotechnology. 2010;**1**(1):3-9. DOI: 10.1504/IJBNN.2010.034121

[53] Almeida JMP, Almeida GFB, Hernandes AC, Mendonça CR. Architecture of lead oxide microcrystals in glass: A laser and etching based method. CrystEngComm. 2016;**31**:5959-5964. DOI: 10.1039/C6CE01255H

[54] Elbossaty WF. Toxicology, biological activity, synthesis, and anti-microbial effects of lead nanoparticles. Journal of Medical Physics and Applied Sciences. 2017;**2**(2:4):1-4. DOI: 10.21767/2574-285X.100011

[55] Ng DQ, Chu Y, Tan SW, Wang SL, Lin YP, Chu CH, et al. In vivo evidence of intestinal lead dissolution from lead dioxide (PbO_2) nanoparticles and resulting bioaccumulation and toxicity in medaka fish. Environmental Science Nano. 2019;**6**(2):580-591. DOI: 10.1039/C8EN00893K

[56] Shamshad M, Soorambail K, Prakash V. Absorption, LD50 and effects of CoO, MgO and PbO nanoparticles on mice "Mus musculus". Journal of Environmental Science, Toxicology and Food Technology. 2015;**9**(2):32-38. DOI: 10.9790/2402-09213238

Section 3

Lead Compounds and Their Properties

Chapter 5

High-Temperature Electrochemical Refining of Secondary Lead

Pavel A. Arkhipov and Yury P. Zaykov

Abstract

The present chapter is devoted to the analysis of the obtained data on the high-temperature electrolytic production of high-purity lead from secondary lead in chloride melts. Kinetic parameters of electrode reactions were calculated, and the sequences of the metal dissolution from the double lead-antimony (Pb-Sb), lead-bismuth (Pb-Bi), and antimony-bismuth (Sb-Bi) alloys were determined. A long-term electrolysis of the antimony (battery scrap), bismuth (lead-bismuth), and lead-containing raw materials in the electrolytic cell of original construction with a porous ceramic diaphragm impregnated with the eutectic $KCl\text{-}PbCl_2$ chloride electrolyte was performed. The anode lead alloy, containing 57.0 wt% of antimony and 36.0 wt% of bismuth, and cathode grade lead were obtained as a result of the electrolysis. The values of lead, antimony, and bismuth separation coefficients were calculated according to the values of the equilibrium potentials of the Pb-Sb, Pb-Bi, and Sb-Bi alloys. The values of separation coefficients were found to be $6.5{\cdot}10^6$–$1.5{\cdot}10^8$ for a single stage at the lead extraction from the Pb-Sb and Pb-Bi alloys, which proves the possibility of a highly effective lead extraction. The value of Sb-Bi alloy separation coefficient ranges from 5.5 to 6.5, which testifies the complexity and low effectiveness of the separation process. An electrolytic refining of lead-bismuth and secondary lead, obtained from the battery scrap, was performed.

Keywords: lead, antimony, bismuth, chloride melt, high-temperature electrolysis, liquid metals, separation coefficient

1. Introduction

Fusible metals play a key role in the development of technologies of separation and extraction of rare and dispersed elements using molten salts [1, 2]. There is published data on cerium extraction using a liquid zinc anode [3], holmium extraction using a liquid bismuth anode [4], and thulium extraction using a liquid zinc anode [5] in the LiCl-KCl electrolyte. Development of the technologies of the electrolytic separation of alloys in molten chloride systems requires understanding of the transformations, which take place at the surface of the liquid metal anode.

Lead is a renewable metal that may be used in materials production multiple times, because its production losses are negligible. A great number of lead-containing devices withdrawn from operation form wastes that require recycling. Apart from that the lead, nonferrous and noble metal-containing cakes, dust, and sludge are formed as side products of nonferrous metallurgical processes. The concentration of the basic admixtures in such lead-containing rare products demonstrates that apart from lead it is possible to extract such products as antimony and bismuth. The technology of electrochemical lead refining in chloride metals is one of the most promising for selective extraction of nonferrous metals.

Electrodissolution of lead-bismuth (Pb-Bi) alloys is described in papers [6–8]. Nichkov studied polarization of the bismuth anode in the molten salt mixtures KCl-NaCl and KCl-LiCl, containing 0.1, 1.0, and 5.0 wt% of $BiCl_3$. The bismuth ionization proceeds mostly with formation of Bi^{3+}. The bismuth current efficiency, which is calculated for the three-electron reaction, is 109%, which denotes the formation of the lowest valence Bi^{+} ions. The increase in the bismuth trichloride concentration in the melt leads to the polarization shift to the regions of more positive potentials, and at $i > 1$ A/cm^2, it reaches the values of −0.76 V relatively to the chloride reference electrode. The measurements of the lead electrode polarization denote that its dissolution takes place at the potentials 0.4 V more negative than those of bismuth dissolution under otherwise equal conditions [6]. Papers [7, 8] illustrated that lead current efficiency is close to 100% if it is calculated for two-electron electrode reactions at the dissolution of Pb-Bi alloys containing up to 70 wt% of Bi. The peculiarities of bismuth and bismuth-containing alloys anode polarization are described in papers [9–13]. Pyatkov [10] measured anode polarization of Sb in the KCl-NaCl electrolyte at the temperatures of 680–780°C. The antimony polarization was found to proceed with formation of ions of various oxidation degree—Sb^{+} and Sb^{3+}. The temperature decrease and the antimony trichloride concentration growth in the melt were found to result in the antimony ionization according to the three-electron electrode reactions [9]. Papers [10, 11] reported on the studies of the possibility of the electrolytic separation of lead-antimony (Pb-Sb) alloys at the temperatures of 973–1073 K in the KCl-NaCl-$PbCl_2$ electrolyte. It is shown that the anode current efficiency is equal to 100% at the 10.0 wt% lead concentration in the anode. As the lead concentration decreases, the anode current efficiency drops to 30–40%. Wei et al. [13] studied the electrochemical behavior of antimony in the LiCl-KCl electrolyte. The square wave function was used to define the number of electrons, which participated in the electrode reaction ($n = 3.02$). This fact denotes that during the experiment, the electrochemical reaction proceeds in a single stage with simultaneous transfer of three electrons. Hirofumi Ebe et al. [14] studied electrochemical transformations of antimony and bismuth in the $AlCl_3$-KCl-NaCl melt with additions of antimony and bismuth trichlorides by voltammetry. The voltammograms of the melts, containing antimony ions, have one peak in the cathode direction and one peak in the anode direction of the potential sweep. The curves of the electrolyte with the bismuth trichloride addition have a more complex form. The authors suggest that under the electrolysis conditions, the oxidation-reduction reaction Bi^{3+}/Bi^{+} at the potentials of 1.1 V takes place (chlorine reference electrode) [13].

The analysis of the literature data on the binary Pb-Sb and Pb-Bi alloys separation elucidates that the process temperature growth leads to the increase in the Sb^{+} and Bi^{+} ion concentrations and therefore to the decrease in the Sb^{3+} and Bi^{3+} ion concentration. The presence of two metal ion types of difference valence decreases the process performance, because part of the current will be spent on the recharge of the ions of the lowest and highest valences. That is why the preferable temperature of the electrochemical separation of double lead alloys should not exceed the melt liquidus temperature by more than 30–100°C.

The present chapter illustrates the results of the studies of the anode dissolution of Pb-Sb, Pb-Bi, and Sb-Bi double alloys.

2. Methodology

Electrodissolution of lead and antimony and lead and bismuth alloys was studied by the galvanostatic method. The polarization was measured in the moment of polarization current switch off [14–17].

The current amplitude was changed from 1 mA to 1A using an IPC-Pro potentiostat (Russian Federation). The polarization value was determined using the IPC-Pro software. The experiments were performed in a quartz cell (**Figure 1**),

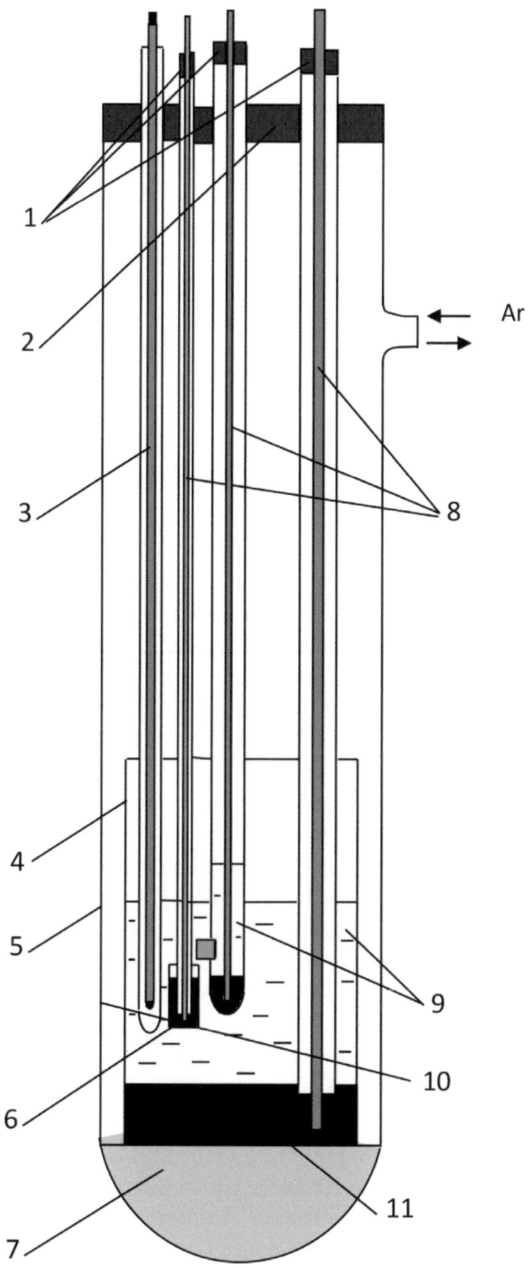

Figure 1.
Schematic of the electrochemical cell. (1) rubber plugs, (2) fluoroplastic lid, (3) thermocouple, (4) alundum crucible, (5) quartz cell, (6) working electrode, (7) platform, (8) current leads to electrodes, (9) electrolyte, (10) reference electrode, (11) auxiliary lead electrode.

which was hermetically closed with a fluoroplastic lid (2), in argon atmosphere. An alundum container (4) was mounted onto a platform made of the fired refractory material (7). A counter electrode (11) and the mixture of potassium and lead chlorides (9) were put into the container. A reference electrode (10), a working electrode, (6) and a thermocouple (3) were lowered into the cell from the gas phase after the electrolyte melting. The alloys of the following compositions were used as working electrodes: lead-antimony, Pb-Sb (70–30), Pb-Sb (50–50), and Pb-Sb (30–70); lead-bismuth, Pb-Bi (59.3–40.7), Pb-Bi (32.5–67.5), Pb-Bi (17.1–82.9), and Pb-Bi (3–97); and antimony-bismuth, Sb-Bi (25–75), Sb-Bi (94–04), and Sb-Bi (99–01). The alloy of the same composition as the working alloy was used as a counter electrode. Capsulated lead served as a reference electrode. Current leads to liquid alloys and to lead were made of molybdenum (8). In the electrolyte area, molybdenum was screened from the melt by alundum. The cell was maintained hermetically sealed with rubber plugs (1). The temperature was controlled using a Varta 207 thermoregulator (Russian Federation). In the electrolyte, the temperature was controlled by a chromel-alumel thermocouple. The electrolyte was prepared from potassium chloride of the chemically pure grade and lead chloride of the analytically pure grade. The prepared electrolyte was electrolytically purified. The compositions of the working electrodes were prepared from lead of the C-1 grade, antimony of the Su-0 grade (antimony concentration 99.9 mol%) and bismuth of the Bi-00 grade (bismuth concentration 99.99 mol%).

3. Measurement results

The polarization of lead-antimony working electrodes Pb-Sb (70–30), Pb-Sb (50–50), and Pb-Sb (30–70) was measured at the temperature of 873 K in the current density region of 0.001–2 A/cm^2. **Figure 2** illustrates in a graphic form the obtained results. The potentials of lead-antimony alloys are appropriately located between the potentials of individual lead and antimony. Lead is the most electronegative element in the system. According to the measurement results, it is seen that ionization proceeds with the insignificant shift of the polarization values to the positive potential region in the samples with the lowest antimony concentration (**Figure 2**, curve 2) at the polarization current amplitude from 0.001 to 0.09 A/cm^2. This is explained by the fact that under experimental conditions, the compositions of salt and metallic phases in the area near the anode layer do not change. As the values of the polarization current exceed 0.1 A/cm^2, the anode potential changes more intensively, but the potentials of pure antimony dissolution are still not reached in the whole interval of current impulses. The increase in the antimony concentration in the anode causes changes in the form of the polarization curves (**Figure 2**, curves 3, 4). The measurement results of the polarization of Pb-Sb alloys containing from 50.0 to 70.0 mol% of antimony demonstrate that the form of the curves does not change. The curve may be conditionally divided into two regions: the changes in polarization of the first region are analogous to the form of polarization curve for alloys with small antimony concentration. As the definite value of i_a is reached, the electrode potential shifts abruptly to the positive polarization region until the corresponding values of antimony ionization are reached. The numerical value of the limiting diffusion current density decreases as the concentration of the more electropositive component of the melt increases.

The polarization of lead-bismuth working electrodes Pb-Bi (59.3–40.7), Pb-Bi (32.5–67.5), Pb-Bi (17.1–82.9), and Pb-Bi (3–97) was measured at the temperature

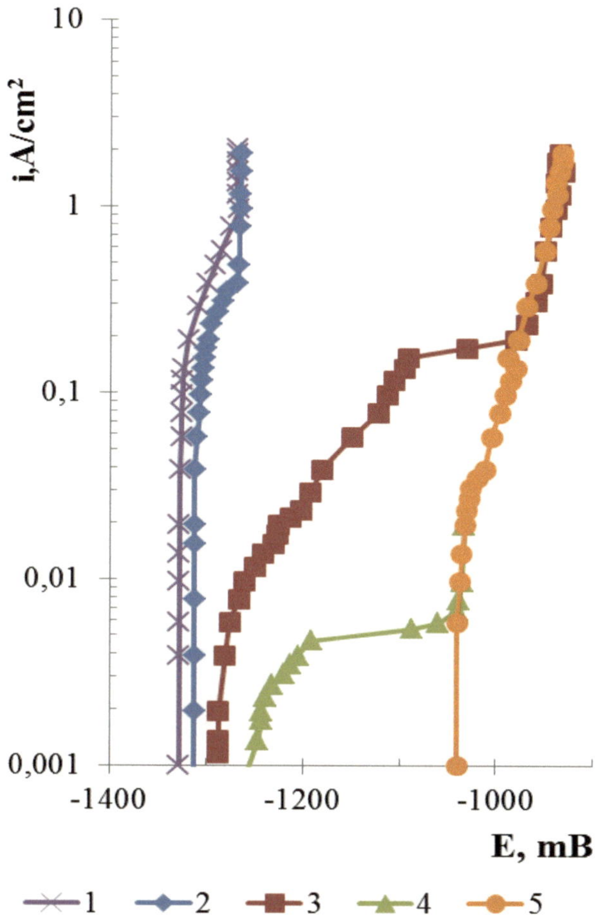

Figure 2.
Polarization of lead-antimony anodes, mol%: 1, Pb; 2, Pb-Sb (70–30); 3, Pb-Sb (50–50); 4, Pb-Sb (30–70); 5, Sb.

of 773 K in the current density interval of 0.001–2 A/cm^2. **Figure 3** illustrates the lead-bismuth electrode polarization measurements.

Analogous to the Pb-Sb alloys, the potentials of the Pb-Bi alloys are located between the potentials of individual lead and bismuth. In addition, the potentials of bismuth are significantly more positive than those of lead. The alloys containing 40.7, 67.5, and 82.9 mol% of bismuth (**Figure 3**, curves 2–4) demonstrate the constancy of measured potentials at the anode current density increase to 0.2 A/cm^2. Within the current density interval of 0.2–2.0 A/cm^2, the anode potential shifts noticeably to the region of electropositive values. The anode potential of the working electrode containing 97 mol% of Bi reaches the electropositive values of bismuth dissolution potentials (**Figure 3**, curve 5). This is the only composition of the working electrode that has the following values.

The polarization of antimony-bismuth working electrodes Sb-Bi (25–75), Sb-Bi (94–04), and Sb-Bi (99–01) was measured at the temperature of 773 K in the current density interval of 0.001–2 A/cm^2. **Figure 4** illustrates the measurement results.

The potentials of the Sb-Bi alloys are located between the potentials of individual antimony and bismuth, and the bismuth potential is more positive than that of

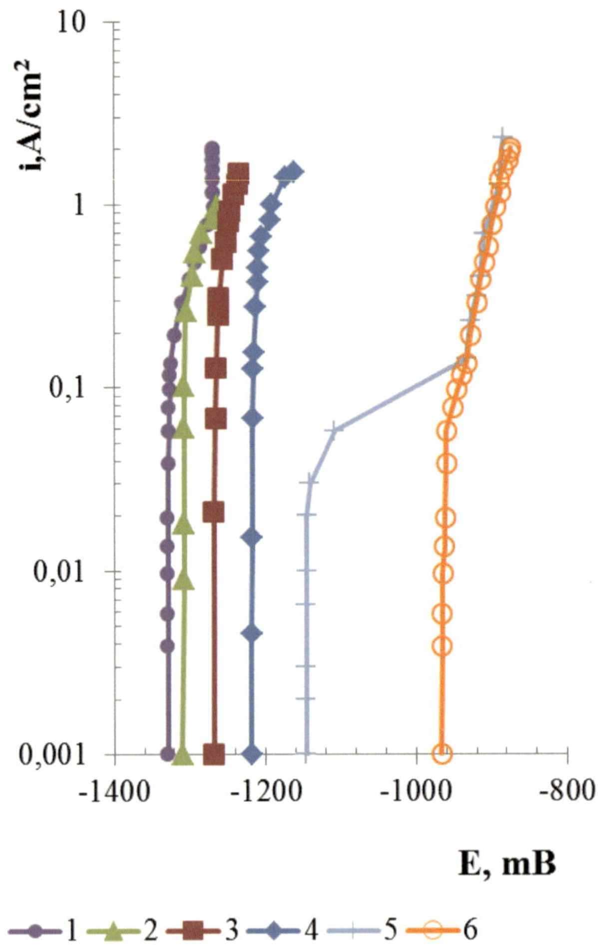

Figure 3.
Polarization of lead-bismuth anodes, mol%: 1, Pb; 2, Pb-Bi (59.3–40.7); 3, Pb-Bi (32.5–67.5); 4, Pb-Bi (17.1–82.9); 5, Pb-Bi (3–97); 6, Bi.

antimony. Polarization curves of all alloys under analysis demonstrate the platform of the limiting diffusion current of antimony ionization. The numerical value of the limiting diffusion current density decreases as the concentration of the more positive alloy component, i.e., bismuth, increases.

The analysis of the general form of the polarization dependencies of double metallic systems allows us to predict that the metal ionization from the alloys proceeds selectively in diffusion regime. The mechanism of dissolution may be described by the following electrode reactions. The most negative metal ionizes on the regions of polarization curves, corresponding to the insignificant deviations of polarization from the alloy equilibrium potentials.

The following electrode reaction describes the dissolution mechanism of the lead-antimony and lead-bismuth electrodes:

$$Pb_{(Pb-Sb,Pb-Bi)} \rightarrow Pb^{2+} + 2e^{-}. \quad (1)$$

This reaction describes the mechanism for the antimony-bismuth alloys:

$$Sb_{(Sb-Bi)} \rightarrow Sb^{3+} + 3e^{-}. \quad (2)$$

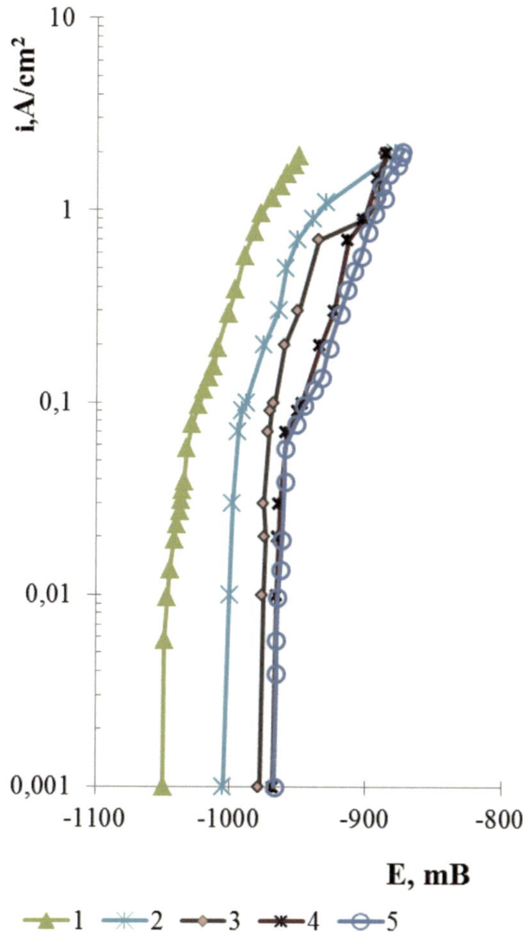

Figure 4.
Polarization of antimony-bismuth anodes, mol%: 1, Sb; 2, Sb-Bi (25–75); 3, Sb-Bi (04–94); 4, Sb-Bi (01–99); 5, Bi (100).

As the value of the polarization current increases, the abrupt shift of the potential to the values corresponding to ionization of more electropositive metal in the alloy is observed. The polarization jump is caused by the lack of the electronegative metal at the surface of the alloy, because the transport rate of the electronegative metal to the reaction spot from the electrode bulk is lower than the ionization rate. This results in the increase in the electropositive metal concentration at the alloy surface. The value of the working electrode potential shifts to the region of positive values, which create the conditions for dissolution of the second component of the alloy.

The following reaction takes place in the lead-antimony electrode:

$$Sb_{(Pb-Sb)} \rightarrow Sb^{3+} + 3e^{-}, \quad (3)$$

and this reaction proceeds in the antimony-bismuth electrode:

$$Bi_{(Pb-Bi,Sb-Bi)} \rightarrow Bi^{3+} + 3e^{-}. \quad (4)$$

The value of the limiting diffusion current of the metal may be used to evaluate the thickness of the diffusion layer in the liquid metal anode [18]:

$$\delta_{Pb-Sb-Bi} = \frac{nFDC_{Me}}{i_{Me}}, \quad (5)$$

where $\delta_{Pb-Sb-Bi}$ is the thickness of the alloy diffusion.
n is the number of electrons.
D is the coefficient of metal atom diffusion in the electrode.
C_{Me} is the concentration of the diffusing atom.
i_{Me} is the limiting diffusion current of metals atoms from the alloy.

The values of lead diffusion coefficients D_{Pb} = 7.67 10^{-5} cm^2/s in the electrode (T = 773 K) and antimony D_{Sb} = 5.3 10^{-5} cm^2/s in the alloy (T = 773 K) are provided in the reference book [19]. Results of the diffusion layer thickness are provided in **Table 1**.

The thickness of diffusion layer in lead liquid alloys with antimony and bismuth of the same order for diffusing lead and antimony is close to the value of diffusion layer in molten salt under self-convection conditions [20]. The presence of clearly marked areas of limiting diffusion currents of ionization of electronegative components in the regions of their small concentrations proves high selectivity of the dissolution of individual elements of the alloys. As the concentration of the electronegative component in the melt (lead or antimony) decreases, the limiting current of its dissolution decreases. Therefore, the limiting current of ionization of the alloy electronegative component may be used to control the alloy separation degree.

The separation of the elements in electrochemical processes with participation of metallic alloys in molten salts was assessed quantitatively on the basis of thermodynamic data for Pb-Sb, Pb-Bi, and Sb-Bi double alloys.

A theoretic possibility to separate elements from the alloys may be evaluated according to the values of equilibrium potentials of alloys of the definite composition in the chloride melt [21]. In our case the Pb-Sb-Bi system in the KCl-$PbCl_2$ melt was used. According to the Nernst equation, the equilibrium potential of the alloy is composed of the value of the standard electrode potential of the potential-determining metal and its ions in the electrolyte ($E^0_{Me^{n+}/Me}$) and a summand, which is determined by the metal ion activity in the electrolyte (a_{Me}) and atom activity in the alloy ($a_{Me(alloy)}$).

The following equation is written for the Pb-Sb alloy:

$$E_P = E^0_{Pb^{2+}/Pb} + \frac{RT}{2F} ln \frac{a_{Pb^{2+}}}{a_{Pb(Pb-Sb)}}. \quad (6)$$

This equation is used for the Pb-Bi alloy:

$$E_P = E^0_{Pb^{2+}/Pb} + \frac{RT}{2F} ln \frac{a_{Pb^{2+}}}{a_{Pb(Pb-Bi)}}. \quad (7)$$

The equilibrium potential of the Sb-Bi alloy is determined as follows:

$$E_P = E^0_{Sb^{3+}/Sb} + \frac{RT}{3F} ln \frac{a_{Sb^{3+}}}{a_{Sb(Sb-Bi)}}. \quad (8)$$

The separation coefficient (θ) is convenient to express the degree of the electrochemical purification of the selective metal dissolution from the alloy. The

Diffusing element	Alloy composition, mol%	C_{Me}, mol/cm^3	i_{Me}, A/cm^2	δ, cm
Pb	Pb-Sb (30.0–70.0)	0.00104	0.05	0.031
Pb	Pb–Bi (3.0–97.0)	0.00165	0.09	0.027
Sb	Sb-Bi (4.0–96.0)	0.00122	0.7	0.018

Table 1.
Thickness of the diffusion layer in liquid alloys at 773 K for double Pb-Sb, Pb-Bi, and Sb-Bi alloys.

separation coefficient is normally written as the quotient of atomic fractions of the separated metals in the electrolyte and in the alloy:

the Pb-Sb alloy:

$$\theta_{Pb/Sb} = \frac{c_{Pb} \cdot c_{Sb^{3+}}}{c_{Pb^{2+}} \cdot c_{Sb}}, \tag{9}$$

the Pb-Bi alloy:

$$\theta_{Pb/Bi} = \frac{c_{Pb} \cdot c_{Bi^{3+}}}{c_{Pb^{2+}} \cdot c_{Bi}}, \tag{10}$$

the Sb-Bi alloy:

$$\theta_{Sb/Bi} = \frac{c_{Sb} \cdot c_{Bi^{3+}}}{c_{Sb^{3+}} \cdot c_{Bi}}. \tag{11}$$

In the majority of cases in the molten salts, when the concentrations of ions of potential-determining metals in the electrolyte have such values that the ion activity coefficients remain constant, the values of standard conditional potentials may be used, at that the equations used to calculate the separation coefficients transform as follows:

for the Pb-Sb alloy:

$$ln\theta_{Pb/Sb} = \frac{-F \cdot E_p + 3F \cdot E^*_{Sb^{3+}/Sb} - 2F \cdot E^*_{Pb^{2+}/Pb}}{RT} + ln\,, \tag{12}$$

for the Pb-Bi alloy:

$$ln\theta_{Pb/Bi} = \frac{-F \cdot E_p + 3F \cdot E^*_{Bi^{3+}/Bi} - 2F \cdot E^*_{Pb^{2+}/Pb}}{RT} + ln\frac{\gamma_{Pb}}{\gamma_{Bi}}, \tag{13}$$

for the Sb-Bi alloy:

$$ln\theta_{Sb/Bi} = \frac{3F \cdot \left(E^*_{Bi^{3+}/Bi} - E^*_{Sb^{3+}/Sb}\right)}{RT} + ln\frac{\gamma_{Sb}}{\gamma_{Bi}}. \tag{14}$$

The equilibrium potentials and element activity coefficients in the alloys of the Pb-Sb and Pb-Bi systems are provided in papers [22, 23]. The antimony and bismuth activity coefficients are reported in paper [24]. The temperature dependencies of the conditional standard electrode potentials of metals are taken from the papers [25–27]. **Figure 5** elucidates calculation results of possible separation coefficients of lead and antimony, of lead and bismuth, as well as of antimony and bismuth according to Eqs. 9–11.

Linear dependencies may be described using the following equations:

the Pb-Sb alloy:

$$ln\theta_{Pb/Bi} = 2.63 + \frac{12547}{T}, R^2 = 0.99 \tag{15}$$

the Pb-Bi alloy:

$$ln\theta_{Pb/Sb} = -0.94 + \frac{14858}{T}, R^2 = 0.99 \tag{16}$$

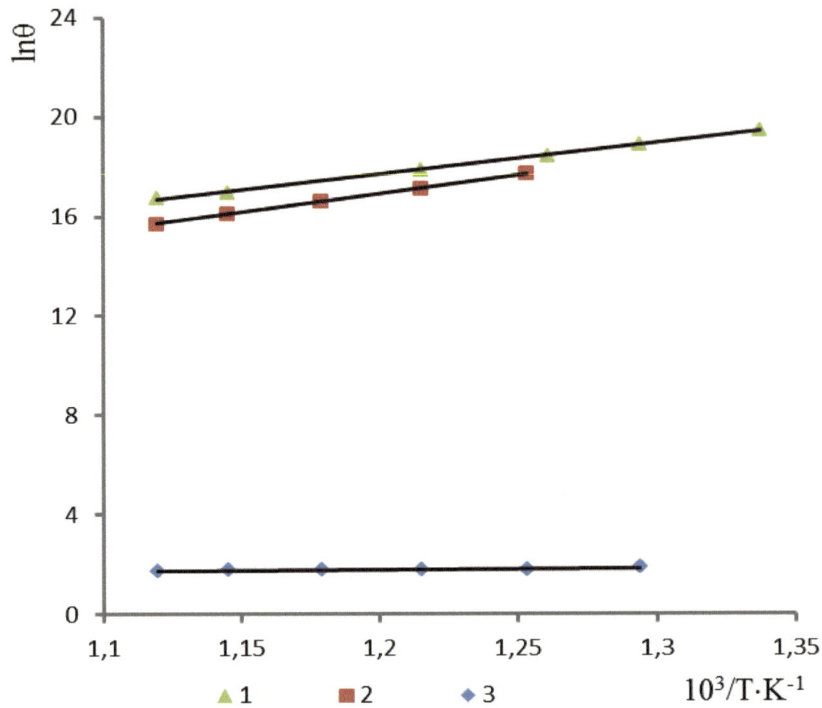

Figure 5.
Dependence of the separation coefficients of metals in double systems on the reverse temperature in the KCl-$PbCl_2$ (50–50 mol%) melt: Pb/Bi; 2, Pb/Sb; 3, Sb/Bi.

the Sb-Bi alloy:

$$ln\theta_{Sb/Bi} = 0.73 + \frac{941}{T}, R^2 = 0.93 \quad (17)$$

The obtained calculation results of the separation coefficients of the double systems according to the experimental values of the equilibrium potentials for the Pb-Sb and Pb-Bi alloys in the temperature range of 748–873 K are $6.5 \cdot 10^6$–$1.5 \cdot 10^8$ for a single stage. The values of separation coefficients allow us to predict a deep selective separation of lead from antimony and bismuth from double alloys. The separation coefficients for the Sb-Bi alloys in the temperature range of 773–873 K are by 6–8 orders of magnitude lower and are located within the range of 5.5 and 6.6. Separation of antimony and bismuth is possible from theoretical point of view if the anode potential is under constant control and the anode current density is readjusted.

The electrolytic cell of original construction was used in the experiments [28].

The experimental electrochemical refining of secondary lead, containing bismuth as a basic admixture, was carried out during 14 days. The range of technological parameters that were tested at the processing of bismuth drosses, which are formed during the lead purification from bismuth by pyrometallurgical method, is as follows:

- Anode current density from 0.2 to 0.35 A/cm^2
- Cathode current density from 0.4 to 0.7 A/ cm^2
- Mass fraction of bismuth in the anode from 2 to 14%

- Current from 300 to 500 A
- Total voltage of the electrolytic cell 8–10 V
- Temperature from 500 to 530°C

Zinc and silver admixtures are always present in the secondary lead. Zinc being the most electronegative metal transfer to the electrolyte first and silver accumulates in the anode metal.

During the electrorefining, lead ions discharge at the cathode to the metallic phase. The bismuth concentration in the cathode metal changes within the range of 0.008–0.011 wt% (**Table 2**) during the whole period of the experiment under the chosen conditions.

Antimony, bismuth, arsenic, and silver accumulate at the anode. The concentration of bismuth at the anode increased from 1 to 13.5 wt% (**Table 3**). In addition,

$i_к$, A/cm^2	t, day	Components concentration, wt%						
		Sb	Sn	Bi	Fe	As	Ag	Zn
0.4	5	0.001	0.0006	0.008	<0.0003	<0.0005	<0.0005	0.001
	6	0.001	0.0006	0.007				0.001
0.5	7	0.001	0.0005	0.008				0.0008
0.6	8	0.001	0.0004	0.008				0.0007
0.7	9	0.001	0.0006	0.009				0.001
	10	0.0008	0.0004	0.008				0.001
	11	0.0006	0.0005	0.009				00009
	12	0.0007	0.0006	0.009				0.006
	13	0.0004	0.0006	0.01				0.001
	14	0.0005	0.0007	0.011				0.001

Table 2.
Lead chemical composition (cathode) at bismuth dross processing.

i_a, A/cm^2	t, day	Components concentration, wt%						
		Sb	Sn	Bi	Fe	As	Ag	Zn
0.2	5	0.022	0.002	5.0	<0.0003	0.001	0.03	0.0005
	6	0.029	0.002	5.5		0.001	0.032	0.0005
0.25	7	0.035	0.002	6.0		0.001	0.038	0.0005
0.3	8	0.41	0.002	6.5		0.001	0.045	0.0005
0.35	9	0.047	0.001	7.0		0.001	0.051	0.0007
	10	0.06	0.002	7.5	0.0030	0.0012	0.057	0.0008
	11	0.08	0.003	8.5	<0.0003	0.0015	0.067	0.0005
	12	0.102	0.002	10.0	<0.0003	0.0017	0.08	0.0007
	13	0.126	0.002	11.5	0.0005	0.0022	0.09	0.0005
	14	0.13	0.0018	13.5	0.0005	0.0041	0.1	0.0005

Table 3.
Chemical composition of the anode metal at bismuth dross processing.

the bismuth concentration at the cathode was by 2–3 orders of magnitude lower, which is in agreement with the theoretical calculations and the polarization studies. The silver concentration increased to 0.011 wt% (**Table 3**).

The zinc concentration does not exceed 0.001 wt% both at the cathode and at the anode. Zinc being the most electronegative metal ionizes first and transfers from the anode to the melt in the form of two valence ions. The zinc ions accumulate in the electrolyte. During the test, the experimental batch of lead of 350 kg was obtained.

The experimental electrochemical refining of secondary lead, containing antimony as a basic admixture alloy, was carried out during 18 days. Black lead, which was obtained by the reducing melting form accumulator scrap, was used as a raw material.

During the electrolytic refining, the antimony concentration in the cathode metal was lower than 0.002 wt% (**Table 4**), and it did not change at the further electrolytic cell operation in the chosen technological regime.

$i_к$, A/cm^2	t, day	Components concentration, wt%						
		Sb	Sn	Bi	Fe	As	Ag	Zn
0.3	2	0.0008	0.0006	0.0005	0.0003	0.0005	0.0003	0.0006
	4	0.0006	0.0006	0.0005	0.0003	0.0005	0.0003	0.0004
0.4	6	0.0007	0.0005	0.001	0.0006	0.0003	0.0003	0.0008
	8	0.0004	0.0004	0.001	0.0006	0.0003	0.0005	0.0007
0.5	10	0.001	0.0006	0.0007	0.0005	0.0005	0.0004	0.0005
	12	0.002	0.0004	0.0005	0.0005	0.0005	0.0005	0.0004
	14	0.002	0.0005	0.0008	0.0003	0.0003	0.0003	0.0009
	16	0.0015	0.0006	0.001	0.0005	0.0005	0.0005	0.0006
	18	0.0011	0.0006	0.001	0.0003	0.0005	0.0005	0.001

Table 4.
Chemical composition of the cathode metal at the processing of black lead obtained by reduction melting from battery scrap.

i_a, A/cm^2	t, day	Components concentration, wt%						
		Sb	Sn	Bi	Fe	As	Ag	Zn
0.3	2	2.3	0.002	0.06	<0.0003	0.03	0.005	0.0008
	4	5.5	0.002	0.13	0.0005	0.08	0.012	0.0003
0.4	6	6.7	0.002	0.16	0.0005	0.10	0.014	0.0007
	8	8.6	0.002	0.20	0.0005	0.14	0.017	0.0005
0.5	10	5.7	0.002	0.13	0.0005	0.09	0.008	0.0008
	12	6.4	0.002	0.14	0.0030	0.10	0009	0.0008
	14	13.2	0.002	0.29	<0.0003	0.22	0.022	0.0006
	16	22.9	0.002	0.36	<0.0003	0.25	0.040	0.0007
	18	26.5	0.002	0.43	0.0025	0.31	0.040	0.0005

Table 5.
Chemical composition of the anode metal at the processing of black lead obtained by reduction melting from battery scrap.

The antimony concentration in the anode metal during the test increased to 26.5 wt%, bismuth concentration increased to 0.43 wt%, arsenic concentration grew to 0.31 wt%, and silver concentration reached 0.04 wt% (**Table 5**).

The experimental batch of cathode lead (250 kg) was obtained.

The test of the electrochemical separation of Pb-Sb and Pb-Bi alloys demonstrated that lead is effectively separated from metal admixtures. The end product, i.e., grade lead, was obtained on the cathode, and alloys of lead with the excess concentrations of bismuth and antimony were obtained on the anode.

4. Conclusions

Anode dissolution of Pb-Sb, Pb-Bi, and Sb-Bi alloys in the $KCl-PbCl_2$ melt at the temperatures of 773–873 K in wide interval of current densities was studied. The anode dissolution of Pb-Sb and Pb-Bi alloys proceed according to the two-electron scheme in the whole interval of current densities (0.07–2.00 A/cm^2), i.e., the anode dissolution of lead takes place. The anode dissolution of Sb-Bi alloys proceeds according to the three-electron scheme in the whole interval of current densities, i.e., the anode dissolution of antimony takes place. The polarization curves illustrate the limiting diffusion current of lead ionization for Pb-Sb and Pb-Bi alloys, and at the Sb-Bi alloy dissolution, the antimony dissolution limiting current is observed.

The thickness of diffusion layer in the liquid metal anode was evaluated according to the limiting diffusion current of the metal. The thickness of the diffusion layers in liquid Pb-Sb, Pb-Bi, and Sb-Bi alloys are of the same order for diffusing lead, and antimony atoms and their values are close to that of the diffusion layer in the molten salt.

The experimentally obtained values of the equilibrium potentials of Pb-Sb, Pb-Bi, and Sb-Bi alloys in the temperature interval of 748–873 K were used to calculate the separation coefficients of double systems. According to the obtained values of separation coefficients, it was found that lead extraction from the Pb-Sb and Pb-Bi alloys is highly effective ($6.5{\cdot}10^6$–$1.5{\cdot}10^8$), whereas the process of bismuth and antimony separation is complex and not effective (5.5–6.5).

Author details

Pavel A. Arkhipov* and Yury P. Zaykov
Institute of High Temperature Electrochemistry of the Ural Branch of Russian Academy of Sciences, Ekaterinburg, Russia

*Address all correspondence to: arh@ihte.uran.ru

References

[1] Agarwal R, Samui P. Enthalpy increment and heat capacity of Pb_3Bi. Journal of Alloys and Compounds. 2010; **508**:333-337

[2] Smolenskii VV, Novoselova AV, Volkovich VA, Osipenko AG, Griffiths TR. Thermodynamics of Nd–Ga–Al and U–Ga–Al alloys and uranium/neodymium separation factor in the molten Ga–Al/3LiCl–2KCl system. Radiochemistry. 2015;**57**(6):591-595

[3] Smolenskiy VV, Novoselova AV, Osipenko AG, Lukyanova YM. The U/La and U/Nd separation coefficients in the Ga-In/3LiCl-2KC1 molten system. Rasplavy. 2015;**1**:49. (in Russian)

[4] Lu W, Liu Y-L, Liu K, Tang S-L, Yuan L-Y, Lu T, et al. Electrochemical extraction of cerium by forming Ce-Zn alloys in LiCI-KCI eutectic on W and liquid Zn electrodes. Journal of the Electrochemical Society. 2015;**162**(9):E179

[5] Han W, Li Z, Li M, Li W, Zhang X, Yang X, et al. Electrochemical extraction of holmium and thermodynamic properties of Ho-Bi alloys in LiCI-KCI eutectic. Journal of the Electrochemical Society. 2017; **164**(4):E62

[6] Li X, Yan Y-D, Zhang M-L, Xue Y, Tang H, Zhou Z-P, et al. $ZnCI_2$ and liquid zinc assisted electrochemical extraction of thulium from LiCI-KCI melt. Journal of the Electrochemical Society. 2014;**161**(5):D248

[7] Ichkov IF, Dmitriyev VY, Raspopin SP. Anode dissolution of bismuth, thorium and lead alloys in molten chloride salts. Izvestiya Vuzov Tsvetnaya Metallurgiya. 1961;**2**:81-87. (in Russian)

[8] Delimarskyi YK, Turov PP, Gitman YB. Electrochemical separation of double alloys of lead with bismuth, antimony, arsenic and tin in a molten electrolyte. Ukrainskiy Khimicheskiy Zhurnal. 1955;**21**:687. (in Russian)

[9] Alabyshev AF, Guelman YM. Electrochemical separation of lead and bismuth in molten electrolyte. Tsvetnye Metally. 1946;**1946**(2):37-43. (in Russian)

[10] Pyatkov VI, Klimovskiy NM, Isovskiy VV, Lebedev VA. Anode dissolution of antimony and its alloys with zinc in chloride-fluoride melts. News of higher educational institutions. Non-Ferrous Metallurgy. 1976;**1976**(1): 82-84. (in Russian)

[11] Sazhin NP, Krol NP, Ilchenko VV. Pyrometallurgical methods of antimony refining. “Scientific works” (Giredmet), M.: Metallurgizdat. 1959;**1**:809. (in Russian)

[12] Rozlovskiy AA, Buldakov AA, Yefimov GN. Collection “Electrochemical Refining of Heavy Fusible Metals from Molten Salts”. Kiev: Naukova dumka; 1971. p. 157. (in Russian)

[13] Wei S, Zhang M, Han W, Yan Y, Xue Y, Zhang M, et al. Electrochemical behavior of antimony and electrodeposition of Mg-Li-Sb alloys from chloride melts. Electrochimica Acta. 2011;**56**:4159-4166

[14] Ebe H, Ueda M, Ohtsuka T. Electrodeposition of Sb, Bi, Te, and their alloys in $AlCl_3$-NaCl-KCl molten salt. Electrochimica Acta. 2007;**53**:100-105

[15] Arkhipov PA, Kholkina AS, Khalimullina YR, Zaykov YP. Anode processes on the liquid metal electrode from the lead alloys. Trudy Kolskogo Nauchnogo Centra. Khimiya i Materialovedeniye. 2018;**2018**(2): 386-388. (in Russian)

[16] Zaykov YP, Arkhipov PA, Plekhanov KA, Ashikhin VV,

Khalimullina YR, Khramov AP. Anode dissolution of Pb-Sb alloys in equimolar mixture of potassium and lead chlorides. Izvestiya Vuzov Tsvetnaya Metallurgiya. 2008;**2008**(4):11-18. (in Russian)

[17] Khalimullina YR, Lebed AB, Ashikhin VV, Zaikov YP, Arkhipov PA, Efremov AN. Anodic behavior of Pb-Sb alloys in chloride melts. In: Summaries 1 of 18th International Congress of Chemical and Process Engineering. Prague, 24–28 August. 2008. p. 259

[18] Lebedev VA. Selectivity of Liquid Metal Electrodes in Molten Halides. Chelyabinsk: Metallurgiya, Chelyabinsk section; 1993. p. 232. (in Russian)

[19] Lepinskikh BM, Belousov AA, Bakhvalov SG. In: Vatolin NA, editor. Transport Properties of Metallurgical Melts and Molten Slag. Metallurgiya: Moskva; 1995. p. 649. (in Russian)

[20] Baraboshkin AN. Electrocrystallization of Metals from Molten Salts. Moskva: Nauka; 1976. p. 138. (in Russian)

[21] Lebedev VA. Determination methods of conditional standard potential of alloys and its possible application in electrochemistry. Rasplavy. 1988;**2**(5):59-66. (in Russian)

[22] Arkhipov P, Khalimullina Y, Zaikov Y, Pershin P. The Study of the Interaction between Pb-Bi Alloys and Chloride Melts (Conference Paper). ISBN: 978-394027637. Proceedings - European Metallurgical Conference, EMC. Vol. 22011. pp. 357-362

[23] Khalimullina YR, Zaikov YP, Arkhipov PA, Ashikhin VV, Скопов ГВ, Kholkina AS, et al. Equilibrium potentials of Pb-Bi alloys in the KCl-$PbCl_2$ melt. Rasplavy. 2010;**2010**(5): 34-43. (in Russian)

[24] Kazhlaeva RI, Shakhtatinskiy MG, Kuliyev AA, Vbigdorovich VN. Study of thermodynamic properties of solutions of the bismuth-antimony system. Zhurnal Fizicheskoi Khimii. 1970;**1970** (10):2460-2464. (in Russian)

[25] Stepanov YA, Kholkina AS, Arkhipov PA, Zaykov YP, Molchanova NG. Equilibrium potentials of ternary alloys Pb-(Sb-Bi) in chloride melts. XVI Rossiyskaya Konferentsiya po fizicheskoy khimii I electrokhimii rasplavlennykh i tverdykh electrolytiv. Ekatherinburg. 2013;**2013**:397-398. (in Russian)

[26] Arkhipov PA, Kholkina AS, Zaikov YP, Molchanova NG. Electrode potentials of bismuth in the mixture of potassium and lead chlorides. Russian Journal of Electrochemistry. 2018;**54**(9): 672-675

[27] Arkhipov PA, Kholkina AS, Zaikov YP, Molchanova NG. Electrode potentials of antimony in the mixture of potassium and lead chlorides. Russian Journal of Electrochemistry. 2020;**56**(1): 82-86

[28] Arkhipov PA, Zaykov YP, Khalimullina YR. Electrolyzer for thin layer metallic lead electrolytic refining. Pat. RF No 2522920, issued 21.05.2014 (in Russian)